Great Transition
The Promise and Lure of the Times Ahead

PAUL RASKIN, TARIQ BANURI, GILBERTO GALLOPÍN,
PABLO GUTMAN, AL HAMMOND, ROBERT KATES, ROB SWART

A report of the *Global Scenario Group*

STOCKHOLM
ENVIRONMENT
INSTITUTE

Global
Scenario
Group

Stockholm Environment Institute - Boston
Tellus Institute
11 Arlington Street
Boston, MA 02116
Phone: 1 617 266 8090
Email: info@tellus.org

SEI Web: http://www.sei.se
GSG Web: http://www.gsg.org
SEI PoleStar Series Report no. 10

Cover:

   Stephen S. Bernow

   Devera Ehrenberg

ISBN: 0-9712418-1-3

♻ Printed on recycled paper

To our grandparents, who labored and dreamed for us.
To grandchildren the world over, for whom we labor and dream.

Table of Contents

Acknowledgements vii

Preface. ix

1. Where Are We?   **1**

    Historical Transitions. 1
    The Planetary Phase. 5
    Branch Point 8

2. Where Are We Headed?   **13**

    Many Futures 13
    Global Scenarios 14
    Driving Forces 19
    Market-driven Development and its Perils. 22
    Barbarization and the Abyss. 25
    On Utopianism and Pragmatism. 28

3. Where Do We Want To Go?   **31**

    Goals for a Sustainable World 31
    Bending the Curve 32
    Limits of the Reform Path 40
    From Sustainability to Desirability 41

4. How Do We Get There?   **47**

    Strategies. 47
    Change Agents. 49
    Dimensions of Transition 54
        Values and Knowledge 55
        Demography and Social Change 57
        Economy and Governance. 60
        Technology and the Environment. 64
    Civilizing Globalization 69

5. History of the Future   71

Prologue 71
Market Euphoria, Interruption and Revival. 73
The Crisis 79
Global Reform. 81
Great Transition. 86
Epilogue 89

6. The Shape of Transition   91

References. 97

Acknowledgements

We are grateful to each of our *Global Scenario Group* colleagues who joined us over the years in an exhilarating exploration of the global past, present and future—Michael Chadwick, Khaled Mohammed Fahmy, Tibor Farago, Nadezhda Gaponenko, Gordon Goodman, Lailai Li, Roger Kasperson, Sam Moyo, Madiodio Niasse, H.W.O. Okoth-Ogendo, Atiq Rahman, Setijati Sastrapradja, Katsuo Seiki, Nicholas Sonntag and Veerle Vandeweerd. This essay is a manifestation of that joint effort.

We thank the Stockholm Environment Institute, Rockefeller Foundation, the Nippon Foundation, and the United Nations Environment Programme for major funding for *GSG* activities over the years, and Steven Rockefeller for both inspiration and a grant for the early stages of writing. We are deeply indebted to Eric Kemp-Benedict for invaluable contributions to the research and modeling, Faye Camardo and Pamela Pezzati for rigorous editing and David McAnulty for publication assistance. We appreciate the comments of the many reviewers of early versions of the manuscript, and particularly wish to thank Bert Bolin, Michael Chadwick, David Fromkin, Nadezhda Gaponenko, Gordon Goodman, Roger Kasperson, Lailai Li, Madiodio Niasse, Gus Speth and Philip Sutton.

We hope the product honors the many wellsprings of collective insight that flowed into it. But any remaining errors of fact, lapses in judgment and failures of imagination are the responsibility of the authors alone.

Preface

"The future is always present, as a promise, a lure and a temptation."

—Karl Popper

The global transition has begun—a planetary society will take shape over the coming decades. But its outcome is in question. Current trends set the direction of departure for the journey, not its destination. Depending on how environmental and social conflicts are resolved, global development can branch into dramatically different pathways. On the dark side, it is all too easy to envision a dismal future of impoverished people, cultures and nature. Indeed, to many, this ominous possibility seems the most likely. But it is *not* inevitable. Humanity has the power to foresee, to choose and to act. While it may seem improbable, a transition to a future of enriched lives, human solidarity and a healthy planet is possible.

This is the story elaborated in these pages. It is a work of analysis, imagination and engagement. As analysis, it describes the historic roots, current dynamics and future perils of world development. As imagination, it offers narrative accounts of alternative long-range global scenarios, and considers their implications. As engagement, it aims to advance one of these scenarios—*Great Transition*—by identifying strategies, agents for change and values for a new global agenda.

The essay is the culmination of the work of the *Global Scenario Group*, which was convened in 1995 by the Stockholm Environment Institute as a diverse and international body to examine the requirements for a transition to sustainability. Over the years, the

GSG has contributed major scenario assessments for international organizations, and collaborated with colleagues throughout the world. As the third in a trilogy, *Great Transition* builds on the earlier *Branch Points* (Gallopín et al., 1997), which introduced the GSG's scenario framework, and *Bending the Curve* (Raskin et al., 1998), which analyzed the long-term risks and prospects for sustainability within conventional development futures.

It has been two decades since the notion of "sustainable development" entered the lexicon of international jargon, inspiring countless international meetings and even some action. But it is our conviction that the *first wave* of sustainability activity, in progress since the Earth Summit of 1992, is insufficient to alter alarming global developments. A new wave must begin to transcend the palliatives and reforms that until now may have muted the symptoms of unsustainability, but cannot cure the disease. *A new sustainability paradigm* would challenge both the viability and desirability of conventional values, economic structures and social arrangements. It would offer a positive vision of a civilized form of globalization for the whole human family.

This will happen only if key sectors of world society come to understand the nature and the gravity of the challenge, and seize the opportunity to revise their agendas. Four major agents of change, acting synergistically, could drive a new sustainability paradigm. Three are global actors—intergovernmental organizations, transnational corporations and civil society acting through non-governmental organizations and spiritual communities. The fourth is less tangible, but is the critical underlying element—wide public awareness of the need for change and the spread of values that underscore quality of life, human solidarity and environmental sustainability.

Global change is accelerating and contradictions are deepening. New ways of thinking, acting and being are urgently needed. But as surely as necessity is the spur for a *Great Transition*, the historic opportunity to shape an equitable world of peace, freedom and sustainability is the magnet. This is the promise and lure of the twenty-first century.

1. Where Are We?

*E*ach generation understands its historic moment as unique, and its future as rife with novel perils and opportunities. This is as it should be, for history is an unfolding story of change and emergence. Each era is unique—but in unique ways. In our time, the very coordinates through which the historical trajectory moves—time and space—seem transformed. Historical time is accelerating as the pace of technological, environmental and cultural change quickens. Planetary space is shrinking, as the integration of nations and regions into a single Earth system proceeds. Amid the turbulence and uncertainty, many are apprehensive, fearing that humanity will not find a path to a desirable form of global development. But a transition to an inclusive, diverse and ecological planetary society, though it may seem improbable, is still possible.

Historical Transitions

Transitions are ubiquitous in nature. As physical or biological systems develop they tend to evolve gradually within a given state or organization, then enter a period of transformation that is often chaotic and turbulent, and finally emerge in a new state with qualitatively different features. The process of movement from a quasi-stable condition through an interval of rapid change to re-stabilization is illustrated in Figure 1. This broad pattern is found across the spectrum of natural phenomena: the forging of matter in the instant after the big bang, the phase shifts between different states of matter as temperature and pressure change, the epigenesis of individual biological creatures and the evolution of life's diverse forms.

With the emergence of proto-humans some 5 million years ago, and especially *Homo sapiens* about 200,000 years ago, a powerful new factor—cultural development—accelerated the process of change on the planet. Cultural change moves at warp speed relative

Figure 1. *Phases of Transition*

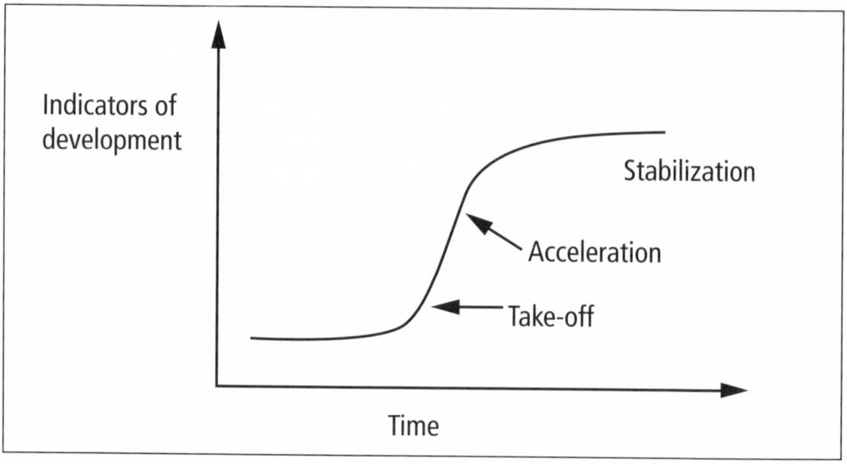

Based on Martens et al. (2001)

to the gradual processes of biological evolution and the still slower processes of geophysical change. A new phenomenon—human history—entered the scene in which innovation and cultural information, the DNA of evolving societies, drove a cumulative and accelerating process of development. With the advent of historical time came a new type of transition, that between the phases of human history that demarcate important transformations in knowledge, technology and the organization of society.

Naturally, the course of history is not neatly organized into idealized transitions. Real history is an intricate and irregular process conditioned by specific local factors, serendipity and volition. The historic record may be organized in different ways, with alternative demarcations between important periods. Yet, a long view of the broad contours of the human experience reveals two sweeping macro-transformations—from Stone Age culture to Early Civilization roughly 10,000 years ago, and from Early Civilization to the Modern Era over the last millennium (Fromkin, 1998). We are now in the midst of a third significant transition, we argue, toward what we shall refer to as the *Planetary Phase of civilization.*

Historical transitions are complex junctures, in which the entire cultural matrix and the relationship of humanity to nature are transformed. At critical thresholds, gradual processes of change working across multiple dimensions—technology, consciousness and institutions—reinforce and amplify. The structure of the socio-ecological system stabilizes in a revised state where new dynamics drive the continuing process of change. But not for all. Change radiates from centers of novelty only gradually through the mechanisms of conquest, emulation and assimilation. Earlier historical eras survive in places that are physically remote and culturally isolated. The world system today overlays an emergent planetary dynamism onto modern, pre-modern and even remnants of Stone Age culture.

Three critical and interacting aspects at each stage are the form of social organization, the character of the economic system, and the capacity for communication. Novel features for each of these dimensions are shown for four historical eras in Table 1.

Table 1. Characteristics of Historical Eras

	Stone Age	Early Civilization	Modern Era	Planetary Phase
Organization	Tribe/village	City-state, kingdom	Nation-state	Global governance
Economy	Hunting and gathering	Settled agriculture	Industrial system	Globalization
Communications	Language	Writing	Printing	Internet

In the Stone Age, social organization was at the tribal and village level, the economy was based on hunting and gathering, and human communication was advanced through the evolution of language. In Early Civilization, political organization moved to the level of the city-state and kingdom, the basis of economic diversification was the surplus generated by settled agriculture, and communication leapt forward with the advent of writing. In the Modern Era, political organization was dominated by the nation-state, the economy became capitalist with the industrial revolution its apotheosis, and communication was democratized through printing. Extending this typology to the Planetary Phase, emerging political, economic and communications features are, respectively, global governance, globalization of the world economy, and the information revolution.

Numerous additional dimensions could be added to characterize the differences in historical eras, such as changing features of art, science, transportation, values, war and so on. But the schematic of Table 1 at least suggests how various aspects of the socio-economic nexus cohere at different stages in the process of historical evolution. In the transition from one coherent formation to another, each of the dimensions transforms. We can follow this process by looking across the rows of the table. Social organization becomes more extensive—tribal, city-state, nation-state and global governance. The economy becomes more diversified—hunting and gathering, settled agriculture, industrial production and globalization. Communications technology becomes more powerful—language, writing, printing, and the information and communication revolution of the current phase.

Societal complexity—the number of variables needed to describe roles, relationships and connectedness—increases in the course of these transitions. Each phase absorbs and transforms its antecedents, adding social and technological complexity. In a heartbeat of geological time, the scale of organization moves from the tribe to the globe, the economy becomes increasingly differentiated, and the technology of communication develops from the capacity for language to the Internet.

Not only does social complexity and the extent of spatial connectedness increase from one epoch to the next, so does the pace of change. Just as historical transitions occur more rapidly than natural evolutionary transitions, historical transitions are accelerating. This is illustrated in Figure 2, which represents schematically the evolution of complexity of the four major historical phases. Since the time-axis is logarithmic, the repetitive pattern suggests that change is accelerating in a regular fashion. The duration of successive eras decreases by roughly a factor of ten—the Stone Age lasted roughly 100,000 years, Early Civilization about 10,000 years and the Modern Era

Figure 2. Acceleration of History

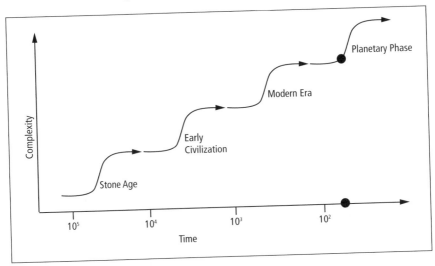

some 1,000 years. Curiously, if the transition to a Planetary Phase takes about 100 years (a reasonable hypothesis, we shall argue) the pattern would continue.

The Planetary Phase

Scanning the broad contours of historical change suggests a long process of increasing social complexity, accelerating change and expanding spatial scale. A premise of much of the contemporary globalization discourse is that humanity is in the midst of a new historical transition with implications no less profound than the emergence of settled agriculture and the industrial system (Harris, 1992). The changing global scene can be viewed through alternative windows of perception—disruption of the planetary environment, economic interdependence, revolution in information technology, increasing hegemony of dominant cultural paradigms and new social and geopolitical fissures.

Globalization is each of these and all of these, and cannot be reduced to any single phenomenon. It is a unitary phenomenon with an array of reinforcing economic, cultural, technological, social and environmental aspects. At the root of the diverse discourse and debate on globalization, and transcending the differences between those who celebrate it and those who resist it, one theme is common. The hallmark of our time is that the increasing complexity and scale of the human project has reached a planetary scale.

Of course human activity has always transformed the earth system to some extent, and the tentacles of global connectedness reach back to the great migrations out of Africa, to the spread of the great religions, and to the great voyages, colonialism and incipient international markets of a century ago. Capitalism has had periods of rapid expansion and integration of regions on the periphery of world markets. It has also had phases of retraction and stagnation associated with economic, political and military crises. The international system and its institutions have been restructured and dominant nations have been displaced (Sunkel, 2001; Ferrer, 1996; Maddison, 1991). At the end of the nineteenth century, the international integration of finance, trade and investment was comparable to contemporary levels when taken as a percentage of the much smaller world economy.

The claim that a planetary phase of civilization is taking shape does not deny the importance of economic expansion and interdependence in earlier eras. Indeed, the increasing imprint of human activity on nature and the expanding reach of dominant nations were necessary antecedents of globalization. The essence of the premise of a planetary transition is that the transformation of nature and the interconnectedness of human affairs has reached a qualitatively new stage. Growing human population and economies inevitably must butt against the resource limits of a finite planet. The increasing complexity and extent of society over hundreds of millennia must at some point reach the scale of the planet itself. That point is now.

Planetary dynamics operating at global scales increasingly govern and transform the components of the earth system. Global climate change influences local hydrology, ecosystems and weather. Globally connected information and communication technology penetrate to the furthest outposts, changing values and cultures, while triggering traditionalist backlash. New global governance mechanisms, such as the World Trade Organization (WTO) and international banks, begin to supersede the prerogatives of the nation-state. The stability of the global economy becomes subject to regional financial disruptions. Excluded, marginalized and inundated with images of affluence, the global poor seek immigration and a better global bargain. A complex mix of despair and fundamentalist reaction feeds the globalization of terrorism. All of these are signs that we have entered a new planetary phase of civilization.

These phenomena are the legacy of the Modern Era of the last thousand years, which brought us to the threshold of planetary society. From the first flickering of the humanistic sensibility nearly a thousand years ago, through the intellectual and theological upheaval of the scientific revolution, to the firestorm of capitalist expansion, modernism challenged the authority of received wisdom, the paralysis of birth-right and class rigidity, and the economic stasis of traditionalism. The culmination was the Industrial Revolution of the last two centuries. It fused a host of modern developments—law-governed institutions, market economies and scientific ingenuity—and tapped into the human potential for accumulation, acquisition and innovation. A permanent revolution in technology, culture and desire spawned an explosion of population, production and economic complexity. Ever hungry for new markets, resources and investment opportunities, the self-expanding and colonizing industrial system began its long march toward a world system.

The world has now entered the Planetary Phase, the culmination of the accelerating change and expansion of the Modern Era. A global system is taking shape with fundamental differences from previous phases of history. We would search in vain for a precise moment that demarcates the origin of the new era. The past infuses the present. Surely the growth of world trade a hundred years ago, the two world wars of the twentieth century and the establishment of the United Nations in 1948 were early signals.

But the primary phenomena that constitute globalization emerged as a cluster over the last two decades. Critical developments between 1980 and the present are seen in:

- The global environment. The world becomes aware of climate change, the ozone hole and threats to biodiversity, and holds its first Earth Summit.
- Technology. The personal computer appears at the beginning of the period and the Internet at the end. A manifold communications and information revolution is launched and biotechnology is commercialized for global markets.
- Geo-politics. The USSR collapses, the Cold War ends and a major barrier to a hegemonic world capitalist system is removed. New concerns appear on the geo-political agenda including environmental security, rogue states and global crime and terrorism.
- Economic integration. All markets—commodity, finance, labor and consumer—are increasingly globalized.
- Institutions. New global actors, such as the WTO, transnational corporations and an internationally connected civil society—and global terrorists, the dialectical negation of planetary modernism—become prominent.

Our hypothesis is that these various elements represent constituent aspects of the global transition. This is illustrated in Figure 3, which shows global connectivity, loosely defined, as following the characteristic S-shaped curve of transition, with "take off" over the last two decades. The schematic suggests that we are in the early phase of an accelerating transition. In this turbulent period, the character of the global system that will emerge from the transition cannot be predicted. The ultimate shape of things to come depends to a great extent on human choices yet to be made and actions yet to be taken.

Branch Point

A transition toward a planetary phase of civilization has been launched, but not yet completed. The critical question is: What form will it take? Inspired by the turn of a new millennium, a stream of popular books, pensive editorials and scholarly essays have sought to understand and find meaning in globalization and

Figure 3. Planetary Transition

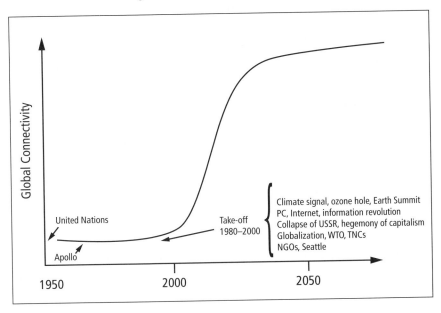

its discontents. The sense that momentous changes are afoot has stimulated a proliferation of explications of what they may portend. As Wittgenstein once noted, the fly in the bottle has difficulty observing the fly in the bottle.

Considerable quantities of old ideological wine have been decanted into the new bottle of global change. As the new realities are refracted through the prism of political and philosophical predilections, the full spectrum of worldviews is revealed—technological optimists and pessimists, market celebrants and Cassandras, social engineers and anarchists. Crudely, archetypal social philosophies can be placed in three broad streams—the evolutionary, the catastrophic and the transformational. They reflect fundamentally different mindsets about how the world works. In the contemporary context, they find expression in divergent outlooks on the long-range prospects for global development.

Evolutionists are optimistic that the dominant patterns we observe today can deliver prosperity, stability and ecological health. Catastrophists fear that deepening social, economic and environmental tensions will not be resolved, with dire consequences for the world's future. Transformationists share these fears, but believe that global transition can be seized as an opportunity to forge a better civilization. In a sense, these represent three different worlds—a world of incremental adjustment, a world of discontinuous cataclysm and a world of structural shift and renewal.

Each worldview sees the future through cloudy crystal balls of interpretation, fear and hope. And in truth, each has a plausible story to tell, for diverse and contradictory forces are at play that could drive global development toward some form of conventional globalization, barbarism or a great historical transition. Fundamentally different worlds could crystallize from the complex and turbulent state of the planet, depending on unfolding events, serendipity and human choice.

Uncertainty and indeterminacy lie deep in the fabric of reality. At the microscopic scale, subatomic matter undergoes discontinuous quantum leaps between states. At the macroscopic scale, as well, apparently identical complex systems can bifurcate into distinct futures at critical crossroads. Similarly, biological systems can absorb and assimilate external disturbances until critical values are exceeded, and then transition to one of multiple possible states. At critical points, small perturbations can have big effects.

Human reflexivity and volition add further dimensions of indeterminacy. The biography of any individual will include decisive moments when experiences and choices shape the lived life, while other possibilities are filed under "what-could-have-been." Human history, too, is not inevitable, as illustrated by counterfactual histories that re-tell the past with plausible "what ifs?" (Ferguson, 1999)—what if Stalin had been ousted in the 1920s or Germany had won World War II? History is a tree of possibilities, in which critical events and decisions are branch points defining one of many alternative pathways.

The horrific terrorist attacks on the United States of September 11, 2001 and their aftermath provide a vivid real-time example of historical branching. "9/11" was a rip in time that defined a "before" and "after," a cultural short-circuit that revealed deep global fissures and interrupted complacent attitudes. At one level, it revealed a strain of pan-Islamic fundamentalism that violently rejects the modernist project itself. As a fanatical fight for purity and against any form of assimilation, it cannot be palliated. At the same time, the despair and anger that is the seedbed for extremism has been brought to the world's attention like never before, exposing the contradictions and failures of global development.

Certainly the world will not be the same after 9/11, but the ultimate implications are indeterminate. One possibility is hopeful: new strategic alliances could be a platform for new multinational engagement on a wide range of political, social and environmental problems. Heightened awareness of global inequities and dangers could support a push for a more equitable form of global development as both a moral and a security imperative. Popular values could eventually shift toward a strong desire for participation, cooperation and global understanding. Another possibility is ominous: an escalating spiral of violence and reaction could amplify cultural and political schisms; the new military and security priorities could weaken democratic institutions, civil liberties and economic opportunity; and people could grow more fearful, intolerant and xenophobic as elites withdraw to their fortresses.

In the critical years ahead, if destabilizing social, political and environmental stresses are addressed, the dream of a culturally rich, inclusive and sustainable world civilization becomes plausible. If they are not, the nightmare of an impoverished, mean and destructive future looms. The rapidity of the planetary transition increases the urgency for vision and action lest we cross thresholds that irreversibly reduce options—a climate discontinuity, locking-in to unsustainable technological choices, and the loss of cultural and biological diversity. Postponing the rectification of how we live together on this planet could foreclose the opportunity for a *Great Transition*.

2. Where Are We Headed?

*I*n the past, new historical eras emerged organically and gradually out of the crises and opportunities presented by the dying epoch. In the planetary transition, reacting to historical circumstance is insufficient. With the knowledge that our actions can endanger the well-being of future generations, humanity faces an unprecedented challenge—to anticipate the unfolding crises, envision alternative futures and make appropriate choices. The question of the future, once a matter for dreamers and philosophers, has moved to the center of the development and scientific agendas.

Many Futures

How do scientific forecasters predict the future of a national economy, local weather or other systems? The key steps are description, analysis and modeling—data are gathered on current conditions, factors are identified that drive change, and future behavior is represented as a set of mathematical variables that evolves smoothly over time. This is a powerful approach when the system under study is well understood and the time horizon is limited. But predictive modeling is inadequate for illuminating the long-range future of our stunningly complex planetary system.

Global futures cannot be predicted due to three types of indeterminacy—ignorance, surprise and volition. First, incomplete information on the current state of the system and the forces governing its dynamics leads to a statistical dispersion over possible future states. Second, even if precise information were available, complex systems are known to exhibit turbulent behavior, extreme sensitivity to initial conditions and branching behaviors at critical thresholds—the possibilities for novelty and emergent phenomena render prediction impossible. Finally, the future is unknowable because it is subject to human choices that have not yet been made.

In the face of such indeterminacy, how can we think about the global future in an organized manner? Scenario analysis offers a means of exploring a variety of long-range alternatives. In the theater, a scenario is a summary of a play. Analogously, development scenarios are stories with a logical plot and narrative about how the future might play out. Scenarios include images of the future—snapshots of the major features of interest at various points in time—and an account of the flow of events leading to such future conditions. Global scenarios draw on both science—our understanding of historical patterns, current conditions and physical and social processes—and the imagination to articulate alternative pathways of development and the environment. While we cannot know what will be, we can tell plausible and interesting stories about what could be.

Rather than prediction, the goal of scenarios is to support informed and rational action by providing insight into the scope of the possible. They illuminate the links between issues, the relationship between global and regional development and the role of human actions in shaping the future. Scenarios can provide a broader perspective than model-based analyses, while at the same time making use of various quantitative tools. The qualitative scenario narrative gives voice to important non-quantifiable aspects such as values, behaviors and institutions. Where modeling offers structure, discipline and rigor, narrative offers texture, richness and insight. The art is in the balance.

Global Scenarios

What global futures could emerge from the turbulent changes shaping our world? To organize thinking, we must reduce the immense range of possibilities to a few stylized story lines that represent the main branches. To that end, we consider three classes of scenarios—*Conventional Worlds, Barbarization* and *Great Transitions*. These scenarios are distinguished by, respectively, essential continuity, fundamental but undesirable social change, and fundamental and favorable social transformation.

Conventional Worlds assume the global system in the twenty-first century evolves without major surprise, sharp discontinuity, or fundamental transformation in the basis of human civilization. The dominant forces and values currently driving globalization shape the future. Incremental market and policy adjustments are able to cope with social, economic and environmental problems as they arise. *Barbarization* foresees the possibilities that these problems are not managed. Instead, they cascade into self-amplifying crises that overwhelm the coping capacity of conventional institutions. Civilization descends into anarchy or tyranny. *Great Transitions*, the focus of this essay, envision profound historical transformations in the fundamental values and organizing principles of society. New values and development paradigms ascend that emphasize the quality of life and material sufficiency, human solidarity and global equity, and affinity with nature and environmental sustainability.

For each of these three scenario classes, we define two variants, for a total of six scenarios. In order to sharpen an important distinction in the contemporary debate, we divide the evolutionary *Conventional Worlds* into *Market Forces* and *Policy Reform*. In *Market Forces*, competitive, open and integrated global markets drive world development. Social and environmental concerns are secondary. By contrast, *Policy Reform* assumes that comprehensive and coordinated government action is initiated for poverty reduction and environmental sustainability. The pessimistic *Barbarization* perspective also is partitioned into two important variants, *Breakdown* and *Fortress World*. In *Breakdown*, conflict and crises spiral out of control and institutions collapse. *Fortress World* features an authoritarian response to the threat of breakdown, as the world divides into a kind of global apartheid with the elite in interconnected, protected enclaves and an impoverished majority outside.

The two *Great Transitions* variants are referred to as *Eco-communalism* and *New Sustainability Paradigm*. *Eco-communalism* is a vision of bio-regionalism, localism, face-to-face democracy and economic autarky. While popular among some environmental and anarchistic subcultures, it is difficult to visualize a plausible path from the globalizing trends of today to *Eco-communalism*, that does not pass through some form of *Barbarization*. In this essay, *Great Transition* is identified with the *New Sustainability Paradigm*, which would change the character of global civilization rather than retreat into localism. It validates global solidarity, cultural cross-fertilization and economic connectedness while seeking a liberatory, humanistic and ecological transition. The six scenario variants are illustrated in Figure 4, which shows rough sketches of the time behavior of each for selected variables.

The scenarios are distinguished by distinct responses to the social and environmental challenges. *Market Forces* relies on the self-correcting logic of competitive markets. *Policy Reform* depends on government action to seek a sustainable future. In *Fortress World* it falls to the armed forces to impose order, protect the environment and prevent a collapse into *Breakdown*. *Great Transitions* envision a sustainable and desirable future emerging from new values, a revised model of development and the active engagement of civil society.

Figure 4. Scenario Structure with Illustrative Patterns

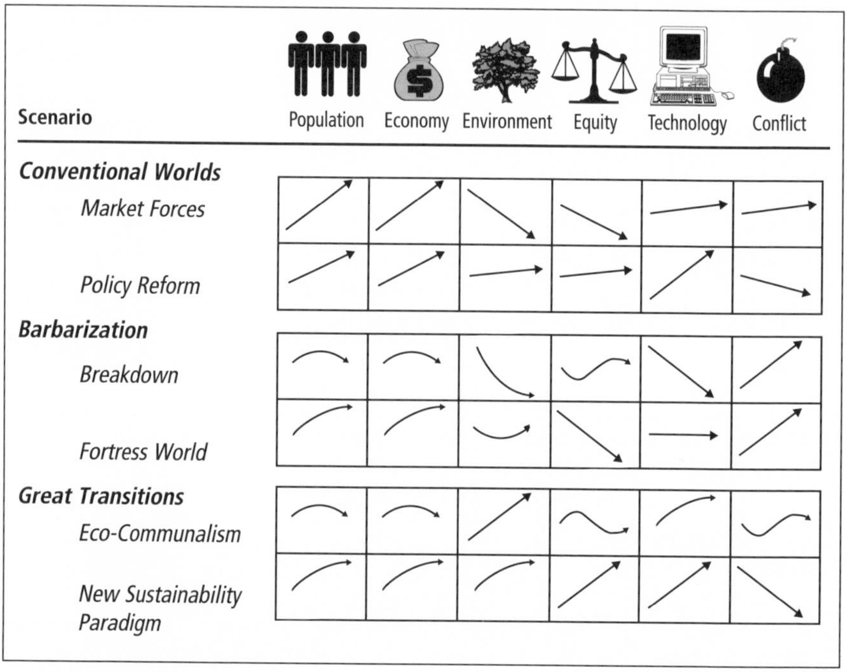

Source: Gallopín et al. (1997)

The premises, values and myths that define these social visions are rooted in the history of ideas (Table 2). The *Market Forces* bias is one of market optimism, the faith that the hidden hand of well-functioning markets is the key to resolving social, economic and environmental problems. An important philosophic antecedent is Adam Smith (1776), while contemporary representatives include many neo-classical economists and free market enthusiasts. In *Policy Reform*, the belief is that markets require strong policy guidance to address inherent tendencies toward economic crisis, social conflict and environmental degradation. John Maynard Keynes, influenced by the Great Depression, is an important predecessor of those who hold that it is necessary to manage capitalism in order to temper its crises (Keynes, 1936). With the agenda expanded to include

Table 2. Archetypal Worldviews

Worldview	Antecedents	Philosophy	Motto
Conventional Worlds *Market*	Smith	Market optimism; hidden & enlightened hand	Don't worry, be happy
Policy Reform	Keynes Brundtland	Policy stewardship	Growth, environment, equity through better technology & management
Barbarization *Breakdown*	Malthus	Existential gloom; population/resource catastrophe	The end is coming
Fortress World	Hobbes	Social chaos; nasty nature of man	Order through strong leaders
Great Transitions *Eco-communalism*	Morris & social utopians Ghandhi	Pastoral romance; human goodness; evil of industrialism	Small is beautiful
New Sustainability Paradigm	Mill	Sustainability as progressive global social evolution	Human solidarity, new values, the art of living
Muddling Through	Your brother-in-law (probably)	No grand philosophies	Que será, será

environmental sustainability and poverty reduction, this is the perspective that underlay the seminal Brundtland Commission report (WCED, 1987) and much of the official discourse since on environment and development.

The dark belief underlying the *Breakdown* variant is that the world faces an unprecedented calamity in which unbridled population and economic growth leads to ecological collapse, rampaging conflict and institutional disintegration. Thomas Malthus (1798), who projected that geometrically increasing population growth would outstrip arithmetically increasing food production, is an influential forerunner of this grim prognosis. Variations on this worldview surface repeatedly in contemporary assessments of the global predicament (Ehrlich, 1968; Meadows et al., 1972; Kaplan, 2000). The *Fortress World* mindset was foreshadowed by the philosophy of Thomas Hobbes (1651), who held a pessimistic view of the nature of man and saw the need for powerful leadership. While it is rare to find modern Hobbesians, many people in their resignation and anguish believe that some kind of a *Fortress World* is the logical outcome of the unattended social polarization and environmental degradation they observe.

The forebears of the *Eco-communalism* belief system lie with the pastoral reaction to industrialization of William Morris and the nineteenth-century social utopians (Thompson, 1993); the small-is-beautiful philosophy of Schumacher (1972); and the traditionalism of Gandhi (1993). This anarchistic vision animates many environmentalists and social visionaries today (Sales 2000; Bossel 1998). The worldview of *New Sustainability Paradigm* has few historical precedents, although John Stuart Mill, the nineteenth century political economist, was prescient in theorizing a post-industrial and post-scarcity social arrangement based on human development rather than material acquisition (Mill, 1848). Indeed, the explication of the new paradigm is the aim of the present treatise.

Another worldview—or more appropriately anti-worldview— is not captured by this typology. Many people, if not most, abjure speculation, subscribing instead to a *Muddling Through* bias, the last row of Table 2 (Lindblom, 1959). This is a diverse coterie, including the unaware, the unconcerned and the unconvinced. They are the passive majority on the grand question of the global future.

Driving Forces

While the global trajectory may branch in very different directions, the point of departure for all scenarios is a set of driving forces and trends that currently condition and change the system:

Demographics

Populations are growing larger, more crowded and older. In typical projections, global population increases by about 50 percent by 2050, with most of the additional three billion people in developing countries. If urbanization trends continue, there will be nearly four billion new city dwellers, posing great challenges for infrastructure development, the environment and social cohesion. Lower fertility rates will lead gradually to an increase in average age and an increase in the pressure on productive populations to support the elderly. A *Great Transition* would accelerate population stabilization, moderate urbanization rates and seek more sustainable settlement patterns.

Economics

Product, financial and labor markets are becoming increasingly integrated and interconnected in a global economy. Advances in information technology and international agreements to liberalize trade have catalyzed the process of globalization. Huge transnational enterprises more and more dominate a planetary marketplace, posing challenges to the traditional prerogatives of the nation-state. Governments face greater difficulty forecasting or controlling financial and economic disruptions as they ripple through an interdependent world economy. This is seen directly in the knock-on effects of regional financial crises, but also indirectly in the impacts of terrorist attacks and health scares, such as mad cow disease in Europe. In a *Great Transition*, social and environmental concerns would be reflected in market-constraining policies, a vigilant civil society would foster more responsible corporate behavior and new values would change consumption and production patterns.

Social Issues

Increasing inequality and persistent poverty characterize the contemporary global scene. As the world grows more affluent for some, life becomes more desperate for those left behind by global economic growth. Economic inequality among nations and within many nations is growing. At the same time, the transition to market-driven development erodes traditional support systems and norms, leading to considerable social dislocation and scope for criminal activity. In some regions, infectious disease and drug-related criminal activity are important social factors affecting development. A central theme of a *Great Transition* is to make good on the commitments in the 1948 Universal Declaration on Human Rights to justice and a decent standard of living for all, in the context of a plural and equitable global development model.

Culture

Globalization, information technology and electronic media foster consumer culture in many societies. This process is both a result and a driver of economic globalization. Ironically, the advance toward a unified global marketplace also triggers nationalist and religious reaction. In their own ways, both globalization, which leaves important decisions affecting the environment and social issues to transnational market actors, and religious fundamentalist reaction to globalization pose challenges to democratic institutions (Barber, 1995). The 9/11 attacks on the United States left no doubt that global terrorism has emerged as a significant driving force in world development. It appears to have contradictory causes—too much modernism and too little. Its hardcore militants seem energized by utopian dreams of a pan-Islamic rejection of Western-oriented global culture. Its mass sympathy seems rooted in the anger and despair of exclusion from opportunity and prosperity. In the clamor for consumerism or its negation, it is sometimes difficult to hear the voices for global solidarity, tolerance and diversity. Yet, they are the harbinger of the ethos that lies at the heart of a *Great Transition*.

Technology

Technology continues to transform the structure of production, the nature of work and the use of leisure time. The continued advance of computer and information technology is at the forefront of the current wave of technological innovation. Also, biotechnology could significantly affect agricultural practices, pharmaceuticals and disease prevention, while raising a host of ethical and environmental issues. Advances in miniaturized technologies could revolutionize medical practices, material science, computer performance and many other applications. A *Great Transition* would shape technological development to promote human fulfillment and environmental sustainability.

Environment

Global environmental degradation is another significant transnational driving force. International concern has grown about human impacts on the atmosphere, land and water resources, the bioaccumulation of toxic substances, species loss and the degradation of ecosystems. The realization that individual countries cannot insulate themselves from global environmental impacts is changing the basis of geo-politics and global governance. A core element of a new sustainability paradigm would be the understanding of humanity as part of the web of life with responsibility for the sustainability of nature.

Governance

There is a significant trend toward democratization and decentralization of authority. On an individual level, there is increased emphasis on "rights," such as women's rights, indigenous rights and human rights broadly conceived. In the private sector, it is reflected in "flatter" corporate structures and decentralized decision-making. Some entities, such as the Internet or NGO networks, have no formal authority structure. The emergence of civil society as an important voice in decision-making is a notable development. A *Great Transition* would see the emergence of a nested governance structure from the local to the global that balances the need to sustain global social and environmental values with the desire for diversity in cultures and strategies.

Market-driven Development and its Perils

In the *Market Forces* scenario, dominant forces and trends continue to shape the character of global development in the coming decades. The tendencies supporting a sustainability transition remain secondary forces. This is the tacit assumption of "business-as-usual" scenarios. But it should be underscored that, like all scenarios, *Market Forces* is a normative vision of the future. Its success requires policy activism, and it will not be easy. Comprehensive initiatives will be required to overcome market barriers, create enabling institutional frameworks and integrate the developing world into the global economic system. This is the program of the IMF, WTO and the so-called "Washington consensus"—we call it the conventional development paradigm.

An earlier study analyzed the *Market Forces* scenario in depth for each global region (Raskin et al., 1998). A thumbnail sketch of selected global indicators is shown in Figure 5. The use of energy, water and other natural resources grows far less rapidly than GDP. This "dematerialization" is due both to structural shifts in the economy—from industry to the less resource-intensive service sector—and to market-induced technological change. But despite such reductions, the pressures on resources and the environment increase as the growth in human activity overwhelms the improved efficiency per unit of activity. The "growth effect" outpaces the "efficiency effect."

Among the projections in the *Market Forces* scenario:

- Between 1995 and 2050, world population increases by more than 50 percent, average income grows over 2.5 times and economic output more than quadruples.
- Food requirements almost double, driven by growth in population and income.
- Nearly a billion people remain hungry as growing populations and continuing inequity in the sharing of wealth counterbalance the poverty-reducing effects of general economic growth.

Figure 5. Global Indicators in *Market Forces* Scenario

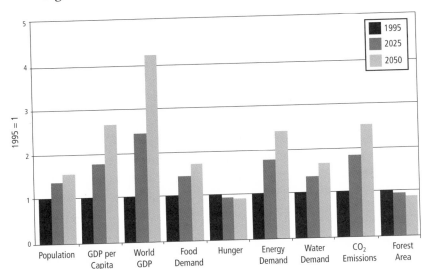

- Developing region economies grow more rapidly than the average, but the absolute difference in incomes between industrialized and other countries increases from an average of about $20,000 per capita now to $55,000 in 2050, as incomes soar in rich countries.
- Requirements for energy and water increase substantially.
- Carbon dioxide emissions continue to grow rapidly, further undermining global climate stability, and risking serious ecological, economic and human health impacts.
- Forests are lost to the expansion of agriculture and human settlement areas and other land-use changes.

A *Market Forces* future would be a risky bequest to our twenty-first century descendants. Such a scenario is not likely to be either sustainable or desirable. Significant environmental and social obstacles lie along this path of development. The combined effects of growth in the number of people, the scale of the economy and the throughput of natural resources increase the pressure that human activity imposes on the environment. Rather than abating, the unsustainable process of environmental degradation that we observe in today's world would intensify. The danger of crossing critical thresholds in global systems would increase, triggering events that could radically transform the planet's climate and ecosystems.

The increasing pressure on natural resources is likely to cause disruption and conflict. Oil would become progressively scarcer in the next few decades, prices would rise and the geopolitics of oil would return as a major theme in international affairs. In many places, rising water demands would generate discord over the allocation of scarce fresh water both within and between countries— and between human uses and ecosystem needs. To feed a richer and larger population, forests and wetlands would continue to be converted to agriculture, and chemical pollution from unsustainable agro-industrial farming practices would pollute rivers and aquifers. Substantial expansion of built-up areas would contribute significantly to land cover changes. The expansion of irrigated farming would be constrained sharply by water shortage and lack of suitable sites. Precious ecosystems—coastal reefs, wetlands, forests and numerous others—would continue to degrade as a result of land change, water degradation and pollution. Increasing climate change is a wild card that could further complicate the provision of adequate water and food, and the preservation of ecosystem goods, services and amenities.

The social and economic stability of a *Market Forces* world would be compromised. A combination of factors—persistence of global poverty, continued inequity among and within nations and degradation of environmental resources—would undermine social cohesion, stimulate migration and weaken international security. *Market Forces* is a precarious basis for a transition to an environmentally sustainable future. It may also be an inconsistent one. The economic costs and social dislocation of increasing environmental impacts could undermine a fundamental premise of the scenario— perpetual global economic growth.

Fraught with such tensions and contradictions, the long-term stability of a *Market Forces* world is certainly not guaranteed. It could persist for many decades, reeling from one environmental, social and security crisis to the next. Perhaps its very instability would spawn powerful and progressive initiatives for a more sustainable and just development vision. But it is also possible that its crises would reinforce, amplify and spiral out of control.

Barbarization and the Abyss

Barbarization scenarios explore the alarming possibility that a *Market Forces* future veers toward a world of conflict in which the moral underpinnings of civilization erode. Such grim scenarios are plausible. For those who are pessimistic about the current drift of world development, they are probable. We explore them to be forewarned, to identify early warning signs and to motivate efforts that counteract the conditions that could initiate them.

The initial driving forces propelling this scenario are the same as for all scenarios. But the momentum for sustainability and a revised development agenda, which seemed so compelling at the close of the twentieth century, collapses. The warning bells—environmental degradation, climate change, social polarization and terrorism—are rung, but not heeded. The conventional paradigm gains ascendancy as the world enters the era of *Market Forces*. But instead of rectifying today's environmental and socio-economic tensions, a multi-dimensional crisis ensues.

As the crisis unfolds, a key uncertainty is the reaction of the remaining powerful institutions—country alliances, transnational corporations, international organizations, armed forces. In the *Breakdown* variant, their response is fragmented as conflict and rivalry amongst them overwhelm all efforts to impose order. In *Fortress World*, powerful regional and international actors comprehend the perilous forces leading to *Breakdown*. They are able to muster a sufficiently organized response to protect their own interests and to create lasting alliances. The forces of order view this as a necessary intervention to prevent the corrosive erosion of wealth, resources and governance systems. The elite retreat to protected enclaves, mostly in historically rich nations, but in favored enclaves in poor nations, as well. A *Fortress World* story is summarized in the box below.

The stability of the *Fortress World* depends on the organizational capacity of the privileged enclaves to maintain control over the disenfranchised. The scenario may contain the seeds of its own destruction, although it could last for decades. A general uprising of the excluded population could overturn the system, especially if rivalry opens fissures in the common front of the dominant strata. The collapse of the *Fortress World* might lead to a *Breakdown* trajectory or to the emergence of a new, more equitable world order.

Fortress World: A Narrative

By 2002, the market euphoria of the last decade of the twentieth-century seems like a naïve and giddy dream. A global economic recession chastens the irrational exuberance of dot-com investors, and the 9/11 terrorist attack awakens a sleepwalking global elite to deep fissures cutting across the geo-political landscape. The nations of the world, mobilized in a cooperative effort to fight terrorism, are offered an unexpected opportunity to redirect development strategy and commit to a form of globalization that is more inclusive, democratic and sustainable. But they do not seize it. The moment of unity and possibility is squandered, in a frenzy of militarism, suspicion and polarization. The empty rhetoric of Earth Summit 2002 is an obituary for the lost era of sustainable development.

Gradually, a coordinated campaign is able to control terrorism at "manageable" levels, although episodic attacks periodically invigorate the politics of fear. The mantra of economic growth, trade liberalization and structural adjustment continues to be heard in the halls of global governance organizations, such as the WTO, the boardrooms of transnational corporations and corridors of national governments. The old ideology of individualism and consumerism flourishes anew, but with a greater respect for the legitimacy of government—as the guarantor of national and individual security, in the first instance, and as an activist partner in enforcing a global market regime, in general.

But it is a bifurcated form of economic globalization limited largely to the so-called "20/20 club"—the 20 percent of nations that are rich and the 20 percent of the elite in nations that are not. The global economy spawns a new class of internationally connected affluent. But there is a counterpoint—the billions of desperately poor whose boats fail to rise with the general economic tide. Some international agencies and some governments continue to mount programs aimed at reducing poverty, promoting entrepreneurship and modernizing institutions. But with financial and political priorities oriented toward security and control, the efforts are woefully inadequate.

As the level of poverty increases and the gulf between rich and poor widens, development aid continues to decline. The remnants of the institutional capacity and moral commitment to global welfare are lost. Meanwhile, environmental conditions deteriorate. Multiple stresses—pollution, climate change, ecosystem degradation—interact and amplify the crisis. Disputes over scarce water resources feed conflict in regions with shared river basins. Environmental degradation, food insecurity and emergent diseases foster a vast health crisis.

(continued)

Fortress World: A Narrative

Tantalized by media images of opulence and dreams of affluence, the excluded billions grow restive. Many seek emigration to affluent centers by any means necessary. Criminal activity thrives in the anarchic conditions, with some powerful global syndicates able to field fearsome fighting units in their battle against international policing activities. A new kind of militant—educated, excluded and angry—fans the flames of discontent. The poison of social polarization deepens. Terrorism resurges, escalating from waves of suicide attacks at popular gatherings and on symbols of globalism, to use of biological and nuclear weapons.

In this atmosphere of deepening social and environmental crisis, conflict feeds off old ethnic, religious and nationalist tensions. Poor countries begin to fragment as civil order collapses and various forms of criminal anarchy fill the vacuum. Even some of the more prosperous nations feel the sting as infrastructure decays and technology fails. The global economy sputters and international institutions weaken, while the bite of climate change and environmental devastation grows fiercer. The affluent minority fears it too will be engulfed by rampant migration, violence and disease. The global crisis spins out of control.

The forces of global order take action. International military, corporate, and governance bodies, supported by the most powerful national governments, form the self-styled Alliance for Global Salvation. Using a revamped United Nations as their platform, a state of planetary emergency is declared. A campaign of overwhelming force, rough justice and draconian police measures sweeps through hot spots of conflict and discontent. With as-needed military and reconstruction support from the Alliance, local forces nearly everywhere are able to subdue resistance and impose stability backed by international peacekeeping units.

A system of global dualism—some call it a *Fortress World*, others Planetary Apartheid—emerges from the crisis. The separate spheres of the haves and have-nots, the included and excluded, are codified in asymmetrical and authoritarian legal and institutional frameworks. The affluent live in protected enclaves in rich nations and in strongholds in poor nations—bubbles of privilege amidst oceans of misery. In the police state outside the fortress, the majority is mired in poverty and denied basic freedoms. The authorities use high-tech surveillance and old-fashioned brutality to control social unrest and migration, and to protect valued environmental resources. The elite have halted barbarism at their gates and enforced a kind of environmental management and uneasy stability.

On Utopianism and Pragmatism

The *Market Forces* worldview embraces both an ambitious vision and a cosmic gamble. The vision is to forge a globally integrated free market by eliminating trade barriers, building market-enabling institutions and spreading the Western model of development. The colossal gamble is that the global market will not succumb to its internal contradictions—planetary environmental degradation, economic instability, social polarization and cultural conflict.

As environments degrade, it is true that some automatic correction acts through the subtle guidance of the "hidden hand" of the market. Environmental scarcity will be reflected in higher prices that reduce demand, and in business opportunities that promote technological innovation and resource substitution. This is why environmental economics draws attention to the critical importance of "internalizing the externalities"—ensuring that the costs of the degradation of environmental resources are monetarized and borne by the producers and consumers who impose such costs. Will such self-correcting mechanisms provide adjustments of sufficient rapidity and scale? To believe so is a matter of faith and optimism with little foundation in scientific analysis or historical experience. There is simply no insurance that the *Market Forces* path would not compromise the future by courting major ecosystem changes and unwelcome surprises.

Another article of faith is that the *Market Forces* development strategy would deliver the social basis for sustainability. The hope is that general economic growth would reduce the ranks of the poor, improve international equity and reduce conflict. But again, the theoretical and empirical foundations for such a salutary expectation are weak. Rather, the national experience in industrial countries over the last two centuries suggests that directed social welfare programs are required to ameliorate the dislocation and impoverishment induced by market-driven development. In this scenario, global poverty would likely persist as population growth and skewed income distributions combine to negate the poverty-reducing effect of growth in average income.

Even if a *Market Forces* future were able to deliver a stable global economic system—itself a highly uncertain hypothesis—the scenario offers no compelling basis for concluding that it would meet the ethical imperatives to pass on a sustainable world to future generations and to sharply reduce human deprivation. Economic and social polarization could compromise social cohesion and make liberal democratic institutions more fragile. Resource and environmental degradation would magnify domestic and international tensions. The unfettered market is important for economic efficiency, but only a fettered market can deliver on sustainability. Environment, equity and development goals are supra-market issues that are best addressed through democratic political processes based on widely shared ethical values and informed by scientific knowledge.

The dream of a *Market Forces* world is the impulse behind the dominant development paradigm of recent years. As the tacit ideology of influential international institutions, politicians and thinkers, it often appears both reasonable and the only game in town. But drifting into the complexity of a global future by relying on such old mind-sets is the sanctuary for the complacent and the sanguine. Ensuring a transition to a sustainable global future requires an alternative constellation of policies, behaviors and values. "Business-as-usual" is a utopian fantasy—forging a new social vision is a pragmatic necessity.

3. Where Do We Want to Go?

*P*ondering the forecaster's question—where are we going?—has led us not to clear answers about the global future, but to disquieting uncertainties. The global trajectory, extrapolated into the future assuming the persistence of dominant trends and values, becomes contradictory and unstable. The curve of development splits into numerous possibilities, with some branches pointing toward barbarous social-scapes and ecological impoverishment. But humans are travelers, not lemmings, who can also ask the traveler's question—where do we want to go? Vision and intentionality is the freedom that draws us forward as surely as the past pushes us onward.

Goals for a Sustainable World

From the tumult of the twentieth century, four great human aspirations crystallized for global society—peace, freedom, material well-being and environmental health. In this century a great transition will need to achieve them.

Peace was to be assured after World War II, but amidst the nuclear arms race, it would be maintained globally but not locally through the long Cold War. The international fight for freedom also began in the late 1940s with the struggle to end imperialism and colonialism, to extend human rights and to combat totalitarian oppression. Then, came a wave of national independence and an international initiative to assist poor countries that aspired to the development standards of the wealthy nations. Lastly, the concern for the well-being of the earth itself emerged in the 1970s, initially focused on natural resources and the human environment, and later extended to the complex systems that support life on Earth.

Now in the early years of the twenty-first century issues of peace and freedom arise again, not only from the many ongoing armed conflicts, but also from acts of terror against non-combatants.

Grappling with these new threats jeopardizes democratic freedoms. The transition beyond war and conflict is part of the sustainability transition. Human rights—economic and social as well as political— need to become universal. Democratic rule, with minority autonomy and rights, needs to be maintained and extended. International conventions already codify many of these goals. For their promise to be fulfilled, they need worldwide ratification and means of enforcement.

The core challenge of development is to meet human needs for food, water and health, and provide opportunities for education, employment and participation. Economically productive and equitable societies can provide literacy, primary and secondary education, and widespread access to advanced education. The end of hunger and deprivation, and the universal right to a healthy and full life are achievable by 2050.

A resilient and productive environment is the precondition for sustaining peace, freedom and development. Preserving the essential health, services and beauties of the earth requires stabilizing the climate at safe levels, sustaining energy, materials and water resources, reducing toxic emissions and maintaining the world's ecosystems and habitats.

At the beginning of a new century, these grand goals for humanity have not been fulfilled, although there has been progress in pursuit of all. The challenge for the future is fashioning a planetary transition that realizes the dream of a more peaceful, free, just and ecologically conscious world.

Bending the Curve

Sustainability goals have been articulated in a long series of formal agreements on human rights, poverty and the environment. But noble sentiments have not been matched by sufficient policy commitments. The vision of sustainability has been a virtual reality superimposed on the real-world push for market globalization.

The broad goals express a powerful ethos for a sustainable world. This is the stirring but intangible music of sustainability. Also needed are the lyrics and the dance—specific targets to concretize the goals and policy actions to achieve them. The *Policy Reform* scenario visualizes how this might occur. The essence of the scenario is the emergence of the political will for gradually bending the curve of development toward a comprehensive set of sustainability targets.

We examined the prospects for a *Policy Reform* future in detail in a previous study (Raskin et al., 1998). The scenario is constructed as a backcast. We begin with a vision of the world in 2025 and 2050 in which minimum sets of environment and social targets have been achieved. We then determine a feasible combination of incremental changes to the *Market Forces* trajectory for meeting these goals. A narrative sketch of a *Policy Reform* scenario is presented in the box below.

What targets are achievable in a *Policy Reform* context? Widely discussed social and environmental objectives provide useful guidance on the scope of the challenge. Naturally, any quantitative targets are provisional, and subject to revision as knowledge expands, events unfold and perspectives change. *Policy Reform* targets for each of the broad sustainability goals—peace, freedom, development and environment—are discussed below and shown graphically in Figure 6, where they are contrasted with patterns in the *Market Forces* scenario.

Peace
The *Policy Reform* path would offer an historic opportunity to address the scourge of war. It seeks an inclusive form of global market development that sharply reduces human destitution, incorporates countries in common international regulatory and legal frameworks and strengthens global governance. The scenario would mitigate underlying drivers of socio-economic, environmental and nationalistic conflict, while adopting international mechanisms for fostering peace and negotiated settlements. In the last decade of the twentieth century, there was an average of 28 major armed conflicts—that is, conflicts that resulted in at least 1,000 battle-related deaths in any single year. The scenario goal is to reduce these to a mere handful by the year 2050.

Freedom

The right of all to participate fully in society without discrimination or bias is a basic right of democratic development. The gradual conferral of equality to women, ethnic groups and racial minorities is a notable achievement of recent decades. The process of eliminating gender and ethnic inequality would accelerate under sustainable development, and could be largely completed by 2050. Figure 6 illustrates this for gender equity as measured by the Gender-Related Development Index that compares life expectancy, educational attainment and income between men and women (UNDP, 2001).

Development

Poverty reduction is the key development goal of the scenario. The incidence of chronic hunger, which now afflicts over 800 million people, is a strong correlate of the poverty nexus. The World Food Summit's call to halve hunger by the year 2015 (FAO, 1996) may have been overly ambitious in light of slow recent progress. The scenario target is to halve hunger by 2025 and halve it again by 2050. Other measures of poverty, such as lack of access to freshwater and illiteracy, have similar patterns of reduction in the scenario. Another useful indicator is average lifespan, which correlates with general human health. With accelerated effort, longevity, which today averages about 60 years in developing countries, could reach 70 years in all countries by 2025, and approach 80 years by 2050.

Environment

Environmental sustainability means reducing human impacts to levels that do not impoverish nature and place future generations at risk. Indicators for climate change, ecosystem loss and freshwater stress are shown in Figure 6.

- The goal for climate change is to stabilize concentrations of greenhouse gases in the atmosphere at safe levels (UNFCCC, 1997). Atmospheric concentrations of carbon dioxide (CO_2), the most important greenhouse gas, have risen from pre-industrial levels of 280 parts per million by volume (ppmv) to about 360 ppmv today. Since the momentum of increasing emissions is inexorable and CO_2 persists in the atmosphere for centuries, climate change cannot be avoided, but it can be moderated. A reasonable, although challenging, goal is to stabilize CO_2 at 450 ppmv by the year 2100. This would keep the cumulative increase in average global temperature below 2°C, a gradual enough change to allow most ecosystems and species to adapt (IPCC, 2001). This will require that greenhouse gas emissions in industrial countries be cut in half over the next 50 years to give "atmospheric space" for poor countries to slowly converge toward common low-emission global standards late in the twenty-first century.

- Climate change is a threat to ecosystems and biodiversity, but not the only one. Land conversions, disruption of freshwater patterns and pollution all contribute. At the least, sustainability requires maintaining sufficient natural areas to ensure adequate protection of ecosystems and associated biodiversity (CBD, 2001; CCD, 2001). Currently, 25 percent of the earth's land is degraded and more than one-fifth of the world's tropical forests have been cleared since 1960 (Watson et al., 1998). A minimum sustainability goal is to halt the loss of ecosystems by 2025 and thereafter begin the process of restoration, a pattern reflected in the targets for forests. While this implies further loss, it is not feasible to completely reverse the tide of destruction in a growing global economy (Raskin et al., 1998).

- Freshwater policy is critical to meeting both environmental and social goals. Today, nearly a third of the world's population is living under moderate or severe water stress (Raskin et al., 1998). As water demands grow, conflict increases in two broad ways—between users in shared river basins and between humanity and nature. The scenario seeks to meet human requirements—the basic needs of people, agriculture and the economy—while maintaining ecosystems. Current trends are not promising—in *Market Forces* the number of people living in water-scarce conditions more than doubles by 2025. A minimum sustainability goal is to moderate water stress through policies to promote water efficiency, waste water recycling and source preservation. Figure 6 shows how water stress could begin to abate with the commitments to water-use efficiency and water resource protection of *Policy Reform*.

Figure 6. Policy Reform and Market Forces Compared: Selected Indicators for Peace, Freedom, Development and Environment

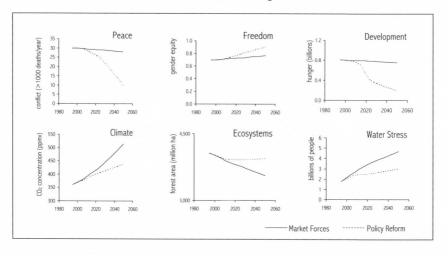

In a *Policy Reform* world, "growth with equity" becomes the prevailing philosophy of development strategies. A host of initiatives increase the incomes of the poor. Reinvigorated multi-national and bi-national livelihood programs build human and institutional capacity. The flow of investment toward the poorest communities and technological transfers accelerate. Market mechanisms for reducing global greenhouse gas emissions and other environmental goals provide additional revenue streams to developing countries, and contribute to the convergence of incomes between developing and industrialized regions. Also, population growth moderates as access to education and effective family planning programs expand.

Relative to unfavorable *Market Forces* trends, the scenario promotes two kinds of equity—between rich and poor countries and within each country. Actions taken to reduce poverty also reduce the immense disparities between the rich and the poor that cleave the current social landscape. Beyond poverty reduction, greater equity in the distribution of wealth between and within countries promotes social cohesion and resilient basis for a peaceful global system. Today the average income in rich countries is nearly seven times that in the rest of the world (and 35 times that in the poorer countries). The scenario reduces this ratio to below 3 by 2050. National equity—defined by the ratio of the incomes of the poorest 20 percent to those of the richest 20 percent, for example—has been declining in many countries. In the *Policy Reform* scenario the drift toward greater inequality is reversed (Raskin et al., 1998).

The environmental goals require substantial decreases in the environmental impacts imposed by rich economies. Elsewhere, impacts increase and then moderate, as poor economies converge toward rich country patterns. On the demand side, the efficiency of energy, water and resource use rapidly increases. On the production side, the transition to renewable energy, ecological agricultural and eco-efficient industrial systems accelerates. *Policy Reform* shows how, with sufficient political commitment, a comprehensive set of policies could begin to redirect development towards sustainability.

These social and environmental initiatives are mutually reinforcing aspects of a unitary project for sustainability. When the poor have access to health care, education and economic security, population growth tends to fall. Poverty reduction helps protect environmental resources, since poverty is both a cause and an effect of environmental degradation. Environmental stability provides the material basis for economic welfare which, in turn, is a precondition for social and economic equity. Greater equity supports cohesion at community, national and global levels. Human solidarity and healthy environments reduce the threat of violence and conflict.

Policy Reform: A Narrative

With the long view of history, globalization stands out as the major theme of the last decades of the twentieth century. Like all turning points, the onset of the planetary phase of world development carries contradictory phenomena in its wake. Superficially, it seems that the dominant engine for change is the rapid advance of a global market system, catalyzed by distance-shrinking transportation and information technology. But a second powerful force, reacting to the predations of heedless global markets, also quietly gestates—the movement for an environmentally sustainable and humane form of development.

The momentum for *Policy Reform* is traced through a series of UN initiatives—the 1972 Stockholm Conference on the Human Environment, the 1987 World Commission on Environment and Development and the 1992 Rio Earth Summit. While these had little immediate effect, in the fullness of time it is clear that they are essential precursors to the remarkable changes of the first decades of the twenty-first century. But it did not seem that way at the time.

Indeed, at the end of the twentieth century, the international momentum for a sustainable future seems squandered. The calls at global conferences for a cohesive agenda for sparing the environment and bringing development to the poor regions of the world appears rarely to go beyond rhetoric to effective action. Special interests squabble, powerful nations resist aligning their development with global environmental goals, and a fragmented system of global governance holds an unending series of topical conferences that offer inspiring but toothless edicts.

But after 2002 history has begun to swing toward sustainable development. A number of factors combine to tilt the balance. The World Summit on Sustainable Development, held in Johannesburg in that year, is a hinge event. The political space for the reform agenda comes in part from the end of market euphoria, so triumphant in the 1990s. At the turn of the new century, a global recession is a reminder that the golden goose of the new prosperity is mortal and that e-commerce has not abolished economic uncertainties. Then the terrorist attacks of 9/11 rip the affluent world from its complacent slumber, at once kindling insecurity, anger and a sense that global development is not working.

Forged in the crucible of a war on terrorism, a new globalism offers an unprecedented opportunity for proactive and cooperative global engagement. The dose of reality persuades government that the internationalization of market opportunities and institutional modernization must proceed on an accelerated basis. The vision at first is confined to delivering on the promise of globalization to assimilate the disaffected and excluded of the earth in the nexus of Western modernism. Free trade institutions are expanded, global governance for the economy is strengthened and international assistance supports a new generation of business and political leaders. At first the vision of an inclusive market-driven world has a salutary effect on the global economy and international security. But the response is insufficient.   *(continued)*

Policy Reform: A Narrative

The environment continues to degrade. The scientific case strengthens that human activity is imperiling global environmental stability. The public grows increasingly impatient, seeing its own evidence in abrupt climate events and mounting reports of species loss. The global economy sputters, and a sense of crisis is amplified by ecological uncertainty and social polarization. In poorer regions, people bitter about the continued failure of globalization to reduce poverty and feeling the bite of climate change demand a new global deal. A combined social, economic and environmental crisis is brewing.

The search begins for a more inclusive, democratic and secure form of development. The world-wide coalition, which began in the fight against global terrorism, extends its mandate to include multilateral action on the environment, arms reduction, international justice and poverty reduction. The goals of international security and sustainable development become interlaced. The media responds and amplifies the mounting environmental and social concerns. NGOs acting through international networks expand their influence. The Internet fuels the global clamor for action. A growing segment of the multinational business community, alarmed at the uncertainties and threats to global stability, become advocates of global policies that reduce risks and provide a level playing field for business.

New political leaders committed to concerted action eventually heed these rising voices. A global consensus emerges on the urgent need for policies to secure environmental resilience and to sharply reduce poverty. The *Policy Reform* response seeks to balance the agendas of those who want no change—*Market Forces* advocates—and those seeking a more fundamental shift in development values—*Great Transition* advocates. The market remains the basic engine for economic growth, supported by trade liberalization, privatization and the global convergence toward the model of development of the rich countries. But globally negotiated targets for environment sustainability and poverty reduction are the basis for constraining and tempering the market. The United Nations is reorganized and its mission refocused on the *Policy Reform* agenda.

The allocation of regional and national responsibilities takes account of the need for rich countries to radically reduce their environmental footprint while assisting poor countries to reduce poverty, to build human capacity and to leapfrog to resource-sparing and environmentally sound technology. The mix of policy instruments for achieving goals—economic reform, regulation, voluntary action, social programs and technology development—varies among regions and nations. Progress toward the global targets is monitored carefully and adjusted periodically. Gradually, global environmental degradation moderates and extreme poverty declines.

Limits of the Reform Path

The *Market Forces* scenario, we have argued, would undermine its own stability by compromising ecological resilience and social coherence. The *Policy Reform* scenario seeks sustainability by constraining market globalization within politically imposed social and environmental targets. But is it enough?

Policy Reform brings both good news and bad news. The good news is that great strides toward a sustainability transition are possible without positing either a social revolution or the *deus ex machina* of a technological miracle. The scenario shows that deep environmental degradation is not a necessary outcome of development. It can be mitigated by new choices for technology, resources and production processes. The cumulative effects of a comprehensive family of feasible incremental adjustments can make a substantial difference. Similarly, poverty and extreme inequity are not inevitable, but result from social policy choices. The long battle against human misery can gradually be won by major actions to promote sustainable livelihoods and greater international and social equity.

The bad news comes in two categories. The first concerns the immense technical challenges of countering conventional development with a reform program. Recall that the *Policy Reform* scenario assumes that the underlying values, lifestyles and economic structures of *Market Forces* endure. *Policy Reform* shows that wise policies on resource efficiency, renewable resources, environmental protection and poverty reduction can, in principle, provide a counter balance. But the required pace and scale of technological and social change is daunting. The reform path to sustainability is like climbing up a down escalator.

The second category of bad news is even more discouraging. The scenario's plausibility rests on a strong postulate—the hypothesis of sufficient political will. For the reform path to succeed, an unprecedented and unyielding governmental commitment to achieving sustainability goals must arise. That commitment must be expressed through effective and comprehensive economic, social and institutional initiatives. But the necessary political will for a reform route to sustainability is today nowhere in sight.

To gain ascendancy, the *Policy Reform* vision must overcome the resistance of special interests, the myopia of narrow outlooks and the inertia of complacency. But the logic of sustainability and the logic of the global market are in tension. The correlation between the accumulation of wealth and the concentration of power erodes the political basis for a transition. The values of consumerism and individualism undermine support for a politics that prioritizes long-range environmental and social well-being. If the dominant interests of popular constituencies and influential power brokers are short-term, politicians will remain focused on the next election, rather than the next generation. It seems that overcoming the dissonance between rhetoric and action will take fundamental changes in popular values, lifestyles and political priorities that transcend *Conventional Worlds* assumptions.

From Sustainability to Desirability

So, *Policy Reform* may not be enough. Taming the juggernaut of conventional globalization with sustainability reforms faces significant technical and political challenges. To these pragmatic concerns about the feasibility of the reform path may be added a normative critique: is it desirable? It envisions a more crowded and engineered global emporium, albeit one where the environment continues to function and fewer people starve. But would it be a place of contentment, choice, and individual and social exploration? It might be a sustainable but undesirable world.

Policy Reform is the realm of necessity—it seeks to minimize environmental and social disruption, while the quality of life remains unexamined. The new sustainability paradigm transcends reform to ask anew the question that Socrates posed long ago: how shall we live? This is the *Great Transitions* path, the realm of desirability.

The new paradigm would revise the concept of progress. Much of human history was dominated by the struggle for survival under harsh and meager conditions. Only in the long journey from tool making to modern technology did human want gradually give way to plenty. Progress meant solving the economic problem of scarcity. Now that problem has been—or rather, could be—solved. The precondition for a new paradigm is the historic possibility of a post-scarcity world

Figure 7. Fulfillment Curve

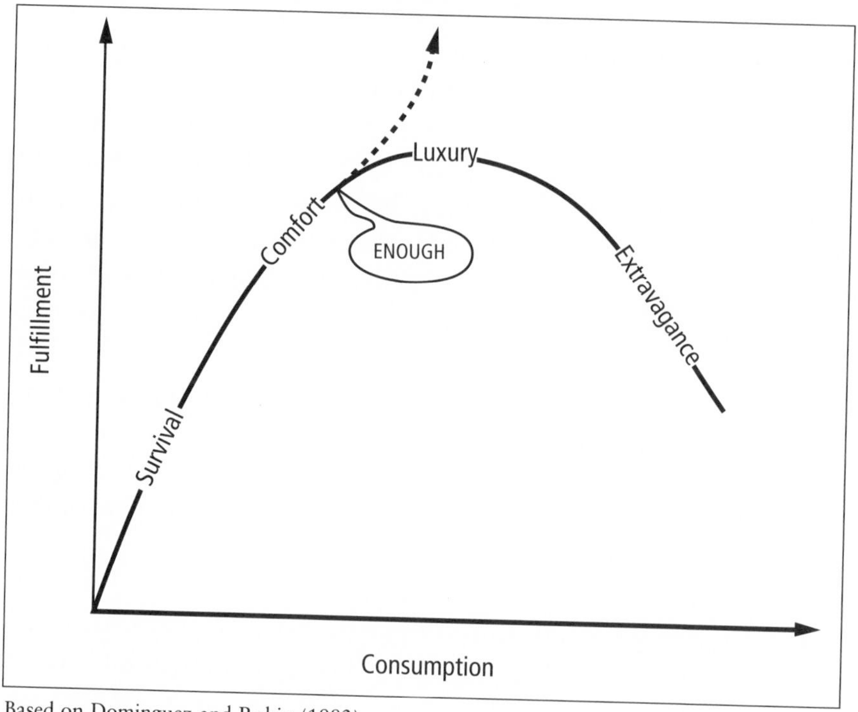

Based on Dominguez and Robin (1992).

where all enjoy a decent standard of living. On that foundation, the quest for material things can abate. The vision of a better life can turn to non-material dimensions of fulfillment—the quality of life, the quality of human solidarity and the quality of the earth. With Keynes (1972), we can dream of a time when "we shall once more value ends above means and prefer the good to the useful."

The compulsion for ever-greater material consumption is the essence of the growth paradigm of conventional worlds. But acquisition as an end in itself can be a substitute for contentment, a hunger that knows no food. The "fulfillment curve" illustrates the erroneous identification of the level of consumption and the quality of life (Figure 7). Past a certain point ("enough"), increased consumption fails to increase fulfillment. Additional costs exceed the marginal satisfaction of additional luxuries as we work to pay for them, learn to use them, maintain and repair them, dispose of them and perhaps feel guilty about having them when others have so little. Profligate consumption sacrifices the cultivation of other aspects of a good life—relationships, creativity, community, nature and spirituality—that can increase fulfillment (the dotted branch in the figure).

A *Great Transition* is galvanized by the search for a deeper basis for human happiness and fulfillment. This has been expressed through diverse cultural traditions. In the new sustainability paradigm, it becomes a central theme of human development. Sustainability is the imperative that pushes the new agenda. Desire for a rich quality of life, strong human ties and a resonant connection to nature is the lure that pulls it toward the future.

Is such a vision possible? It does not seem promising judging by the global scene today, so full of antagonism, inequity and the degradation of nature and the human spirit. Yet, the cunning of history is sure to bring surprises. Some may not be welcome. But favorable possibilities are also plausible.

Later we offer a "history of the future," a hypothetical account of the initial stages of a *Great Transition*. It is written from the perspective of the year 2068 as the transition continues to unfold. What lies beyond this process of change? More change, no doubt. Though an ideal planetary society can never be reached, we can imagine good ones. Distant visions guide the journey. One possibility is sketched in the following box.

A Distant Vision

Here is a civilization of unprecedented freedom, tolerance and decency. The pursuit of meaningful and fulfilling lives is a universal right, the bonds of human solidarity have never been stronger and an ecological sensibility infuses human values. Of course, this is not paradise. Real people live here. Conflict, discontent, mean-spiritedness and tragedy have not been abolished. But during the course of the twenty-first century the historic possibility was seized to redirected development toward a far more sustainable and liberatory world.

The fabric of global society is woven with diverse communities. Some are abuzz with cultural experimentation, political intensity and technical innovation. Others are slow-paced bastions of traditional culture, direct democracy and small-is-beautiful technology. A few combine reflection, craft skill and high esthetics into a kind of "sophisticated simplicity," reminiscent of the Zen art of antiquity. Most are admixtures of countless subcultures. The plurality of ways is deeply cherished for the choice it offers individuals and the richness it offers social life.

The old polarizing dualities—cosmopolitanism versus parochialism, globalism versus nationalism and top-down versus bottom-up—have been transcended. Instead, people enjoy multiple levels of affiliation and loyalty—family, community, region and planetary society. Global communication networks connect the four corners of the world, and translation devices ease language barriers. A global culture of peace and mutual respect anchors social harmony.

The World Union (née the United Nations) unifies regions in a global federation for co-operation, security and sustainability. Governance is conducted through a decentralized web of government, civil society and business nodes, often acting in partnership. Social and environmental goals at each scale define the "boundary conditions" for those nested within it. Subject to these constraints, the freedom to fashion local solutions is considerable—but conditional. Human rights and the rights of other governance units must be respected. While sophisticated conflict resolution processes limit conflict, the World Union's peace force is called on occasion to quell aggression and human rights abuse.

Preferred lifestyles combine material sufficiency and qualitative fulfillment. Conspicuous consumption and glitter are viewed as a vulgar throwback to an earlier era. The pursuit of the well-lived life turns to the quality of existence—creativity, ideas, culture, human relationships and a harmonious relationship with nature. Family life evolves into new extended relationships as population ages and the number of children decreases. People are enriched by voluntary activities that are socially useful and personally rewarding. The distribution of income is maintained within rather narrow bounds. Typically, the income of the wealthiest 20 percent is about two or three times the income of the poorest 20 percent. A minimum guaranteed income provides a comfortable but very basic standard of living. Community spirit is reinforced by heavy reliance on locally produced products, indigenous natural resources and environmental pride.   *(continued)*

A Distant Vision

The economy is understood as the means to these ends, rather than an end in itself. Competitive markets promote production and allocation efficiency. But they are highly fettered markets tamed to conform to non-market goals. The polluter pay principle is applied universally, expressed through eco-taxes, tradable permits, standards and subsidies. Sustainable business practices are the norm, monitored and enforced by a vigilant public. Investment decisions weigh carefully the costs of indirect and long-term ecological impacts. Technology innovation is stimulated by price signals, public preferences, incentives and the creative impulse. The industrial ecology of the new economy is virtually a closed loop of recycled and re-used material, rather than the old throwaway society.

Some "zero growth" communities opt to maximize time for non-market activities. Others have growing economies, but with throughputs limited by sustainability criteria. In the formal economy, robotic production systems liberate people from repetitive, non-creative work. Most everywhere a labor-intensive craft economy rises alongside the high technology base. For the producer, it offers an outlet for creative expression; for the consumer, a breathtaking array of esthetic and useful goods; for all, a rich and diverse world.

Long commutes are a thing of the past. Integrated settlements place home, work, shops and leisure activity in convenient proximity. The town-within-the-city balances human scale community with cosmopolitan cultural intensity. Rural life offers a more serene and bucolic alternative, with digital links maintaining an immediate sense of connectedness to wider communities. Private automobiles are compact and pollution free. They are used in niche situations where walking, biking and public transport options are not available. Larger vehicles are leased for special occasions and touring. Advanced mass transportation systems link communities to local hubs, and those hubs to one another and to large cities.

The transition to a solar economy is complete. Solar cells, wind, modern biomass and flowing water generate power and heat buildings. Solar energy is converted to hydrogen, and used, along with direct electricity, for transportation. Advanced bio-technology is used cautiously for raw materials, agriculture and medicine. Clean production practices have eliminated toxic pollution. Ecological farming makes use of high inputs of knowledge, and low inputs of chemicals to keep yields high and sustainable. Population stabilization, low-meat diets and compact settlements reduce the human footprint, sparing land for nature. Global warming is abating as greenhouse gas emissions return to pre-industrial levels. Ecosystems are restored and endangered species are returning, although scars remain as reminders of past heedlessness.

This is not the end of history. In some sense, it is the beginning. For at last, people live with a deep awareness of their connection to one another, future generations and the web of life.

4. *How Do We Get There?*

*H*ow can we navigate the planetary transition toward a sustainable and desirable global society? *Market Forces* could shipwreck on the shoals of environmental and social crises, and risk sinking into the barbarism of a *Fortress World*. The *Policy Reform* vision would steer toward sustainability with programs for improving technology and reducing poverty, but the momentum of global economic growth could swamp incremental adjustments. And if consumer culture prevails, where would vision and political leadership come from? We must look to more fundamental course changes to guarantee safe passage.

Strategies

The *Great Transitions* approach to a sustainable civilization builds on the wealth-generating features of *Market Forces* and the technological change of *Policy Reform*. But it transcends them by recognizing that market-led adaptations and government-led policy adjustments are not enough. *Great Transitions* adds a third ingredient—a values-led shift toward an alternative global vision. Powerful additional opportunities for mending the global environment and forging more harmonious social conditions would then open. The new development paradigm would include lifestyle changes and greater social solidarity. The distinctions between *Market Forces*, *Policy Reform* and *Great Transitions* visions are illustrated in Figure 8.

Market Forces maintains the conventional correlation of human well-being and the level of consumption, with material consumption, in turn, driving greater throughput of natural resources and impact on the environment. In the *Policy Reform* strategy, the link between well-being and consumption is maintained, but consumption is decoupled from throughput (the "dematerialization wedge"). *Great Transitions* adds a second "lifestyle wedge" that breaks the lockstep connection between consumption and well-being. Environmental

Figure 8. Tools for a Transition

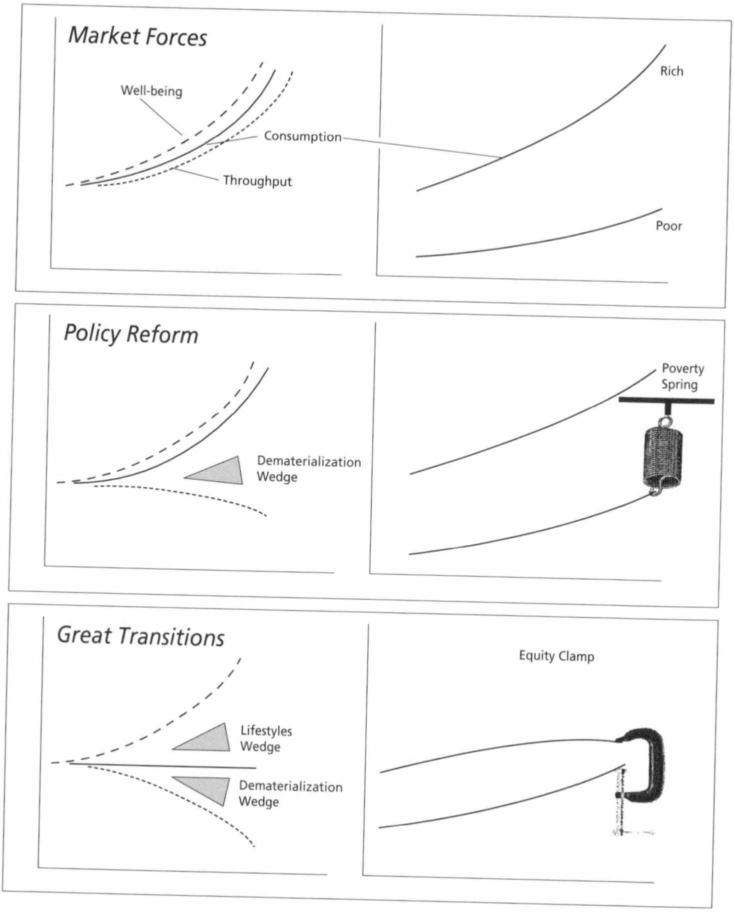

Market Forces

Well-being

Consumption

Throughput

Rich

Poor

Policy Reform

Dematerialization Wedge

Poverty Spring

Great Transitions

Lifestyles Wedge

Dematerialization Wedge

Equity Clamp

Source: "Wedges" based on Robinson and Tinker (1996)

impacts may be decomposed into the product of human activity—miles driven, steel produced, food harvested and so on—and impact per activity. *Policy Reform* focuses on the second factor, introducing efficient, clean and renewable technologies that reduce impacts per activity. *Great Transitions* complements such technology improvements with lifestyles and values changes that reduce and change activity levels in affluent areas, and provide an alternative vision of development globally.

A second critical distinction between the scenarios concerns equity, as illustrated in the right-hand column of sketches in Figure 8. In the *Market Forces* world, the economic growth of the poorer regions of the world is more rapid than the rich regions, but, nevertheless, the absolute difference between rich and poor widens. At the bottom of the income pyramid, a billion people remain mired in absolute poverty. *Policy Reform* strategies substantially reduce absolute poverty through targeted aid and livelihood programs (the "poverty spring"). While the yawning gap between rich and poor is partially abated, global and national inequity remains a threat to social cohesion. Poverty eradication is a fundamental tenet of *Great Transitions*, as well. But in addition to pulling up the bottom, great value is placed on urgently creating more just, harmonious and equitable social relations (the "equity clamp").

Conventional Worlds strategies operate on the direct levers of change that can influence economic patterns, technology, demographics and institutions. Mainstream development policy focuses on these proximate drivers. A *Great Transition* would go deeper to the root causes that shape society and the human experience. These ultimate drivers include values, understanding, power and culture (Figure 9). Proximate drivers are responsive to short-term intervention. The more stable ultimate drivers are subject to gradual cultural and political processes. They define the boundaries for change and the future. The *Great Transition* project would expand the frontier of the possible by altering the basis for human choice.

Change Agents

All global visions inevitably confront the question of agency: Who will change the world? The agents driving the *Market Forces* scenario are global corporations, market-enabling governments and a consumerist public. In *Policy Reform*, the private sector and consumerism remain central, but government takes the lead in aligning markets with environmental and social goals. Civil society and engaged citizens become critical sources of change for the new values of *Great Transitions*.

Figure 9. Proximate and Ultimate Drivers

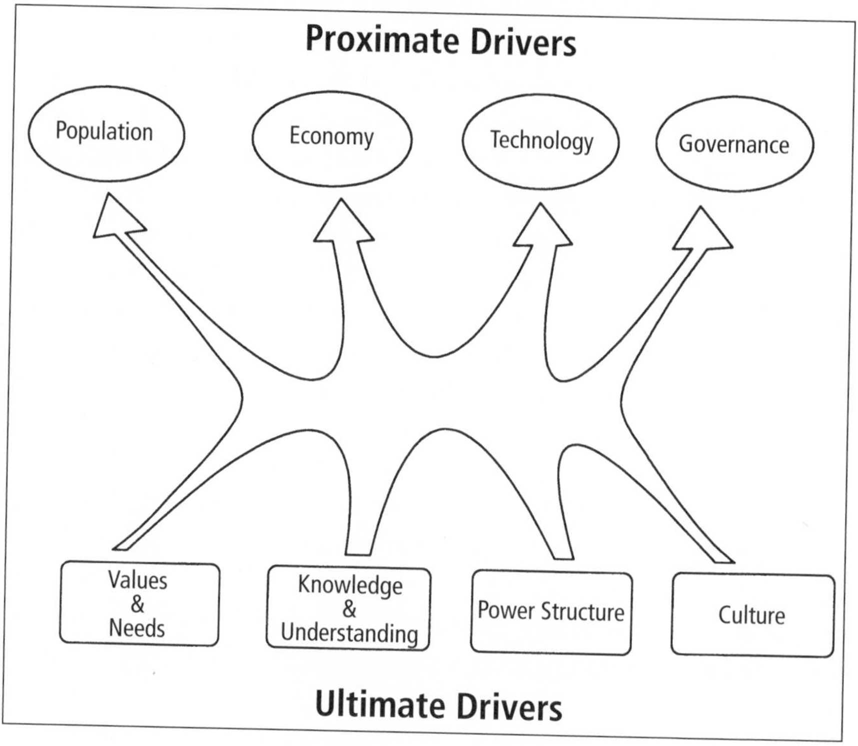

Proximate Drivers

Population   Economy   Technology   Governance

Values & Needs   Knowledge & Understanding   Power Structure   Culture

Ultimate Drivers

In truth, all social actors shape—and are shaped by—world development. The play is difficult to distinguish from the players. The prospects for a *Great Transition* depend on the adaptations of all institutions—government, labor, business, education, media and civil society. But three emerging global actors—intergovernmental organizations, transnational corporations and non-governmental organizations—move to center stage. The fourth essential agency is less tangible—public awareness and values, especially as manifested in youth culture. Meanwhile, other powerful global players—criminal organizations, terrorist rings and special interest groups—lurk in the wings, threatening to steal the show.

The formation of global and regional intergovernmental organizations has tracked the emergence of the Planetary Phase. The United Nations, in particular, embodies the hope that world peace, human rights and sustainable development might rise from the destruction and suffering of the twentieth-century. But the UN has not been given the authority to fulfill that lofty mission, its effectiveness compromised by the politics of its member nation-states. Still, it remains the legitimate collective voice of the world's governments.

That voice would differ across scenarios. In a *Market Forces* world, power moves to the private sector, international banks and the WTO—the UN is largely a toothless platform for international conferences, high-minded rhetoric and crisis management. But in *Policy Reform*, the UN becomes a key venue for implementing environmental and social goals. In *Barbarization* scenarios, the UN is relevant only to historians. In a *Great Transition*, a reorganized UN expresses the international solidarity of the new development agenda as the dominance of the nation-state fades.

To a great extent, the evolution of intergovernmental entities will reflect the political imperatives of the ascendant global system. The ultimate source of the value changes and political choices for a *Great Transition* lie elsewhere. But the UN and the others are not simply dependent variables in the calculus of global change; at critical moments they can provide leadership and initiative for the transition, as well.

The scale, market reach and political might of transnational corporations have soared since World War II. The power of transnational corporations continues to grow in *Market Forces*. *Policy Reform* requires their support, or at least acquiescence; big business comes to understand sustainable development as a necessary condition for preserving the stability of world markets. The *Great Transition* process transforms the role of business. As the new values spread among the consuming public, forward-looking corporations seize the new reality as a business opportunity and a matter of social responsibility. In partnership with government and citizens' groups they establish tough standards for sustainable businesses and innovative practices to meet them.

To some degree, business can drive progressive change. Many win-win opportunities are available for bringing the corporate bottom line of profit into harmony with the societal bottom line of sustainable development. Most directly, good environmental management at facilities can reduce the costs and risks of doing business. In addition, some companies can expand their market share by projecting an image of corporate responsibility. Some business visionaries advocate sustainability as both a business and moral imperative. But the aggregation of these adjustments does not guarantee a transition, nor are business-initiated changes likely to maintain momentum when economic conditions turn sour or the public's interest in sustainability wanes. Nevertheless, sustainability-oriented businesses are an important part of the dynamic of change as they constructively respond to, and reinforce, new pressures from consumers, regulators and the public.

Non-governmental organizations—the organizational expression of civil society—are critical new social actors in global, regional and local arenas (Florini, 2000). The explosive growth in the number and diversity of NGOs has altered the political and cultural landscape. They use modern communications technology to build public awareness and mount campaigns to influence policy and alter corporate behavior. At official international meetings, some are inside the building as active participants, and some are in the streets, challenging the drift of globalization and, in some cases, globalization itself. They are for the most part positive forces for fostering debate and progressive change. But on the dark side, it must be noted, are organizations of terrorists and criminals—perverse forms of NGOs that also use modern information technology, but to spread violence, hate and fear.

NGO success stories include micro credit, social forestry, environmental advocacy, community development and appropriate technology programs. These activities enable communities to participate more effectively in economic and social decisions, and give poor populations access to skills and financial resources. They influence business practices through monitoring, direct action and boycotts. They promote alternative lifestyles. More recently, global public policy networks have begun to link individuals and organizations from multiple countries and stakeholder groups. These networks engage in research, public outreach, advocacy and organized protest on a range of sustainability issues (Reinicke et al., 2000; Banuri et al., 2001).

In doing all this, civil society organizations fill major gaps in public policy-making. By harnessing expert opinion from a diverse set of viewpoints, they have helped create capacity to analyze and respond to emerging problems. By mobilizing stakeholder groups and by refining participatory methodologies, they have helped create channels of public participation. By increasing public awareness, they have fostered transparency in decision-making. Finally, and most importantly, they have injected ethical and normative voices into the political arena.

Like all social actors, civil society is a phenomenon in flux, transformed by the very processes of global change it seeks to influence. Unleashing wellsprings of energy and activism, the new civil society is beginning to discover itself as a globally connected force for change, experimenting with different forms of alliance and networking. Yet, as a global movement, it remains fragmented and responsive, lacking a cohesive positive social vision and coherent strategy.

A critical uncertainty for a *Great Transition* is whether civil society can unify into a coherent force for redirecting global development. This would require a coalescence of seemingly unrelated bottom-up initiatives and diverse global initiatives into a joint project for change. Such a force would entail a common framework of broad principles based on shared values fostered through the activities of educational, spiritual and scientific communities.

Intergovernmental organizations, transnational corporations and civil society are key global actors. The underlying engine of a *Great Transition*, however, is an engaged and aware public, animated by a new suite of values that emphasizes quality of life, human solidarity and environmental sustainability. In this regard, the international youth culture will be a major force for change, albeit a diffuse one. Connected by the styles and attitudes spread by media, global youth represent a huge demographic cohort whose values and behaviors will influence the culture of the future. If they evolve toward consumerism, individualism and nihilism, the prospects would not be promising. But as globalization and its problems mature, the world's youth could rediscover idealism in a common project to forge a *Great Transition*.

Finally, it should be noted that some see technology, rather than social agents, as the primary driver of change. Optimists celebrate the potential for information technology, biotechnology and artificial intelligence to entrain a broad web of favorable societal transformation. Pessimists warn of a dehumanized digital, robotic and bio-engineered society. But all scenarios—*Market Forces, Policy Reform, Great Transitions* and even *Fortress World*—are compatible with the continuing technological revolution. Technology is not an autonomous force. The agenda, pace and purpose of innovation is shaped by the institutions, power structure and choices of society.

To envision a *Great Transition* is to imagine the continued evolution of civil society organizations toward formalization and legitimacy, new roles for business and government and, especially, new values and participation by global citizens. With no blueprint, this will be a long project of social learning and discovery, a process of experimentation and adaptation (BSD, 1998). Where political will is lacking, *civil will* drives the transition forward. The question is whether change agents will remain fractional and fragmented, or whether they will expand and unify to realize the historic potential for transformation. If the many voices form a global chorus, it will herald a new sustainability paradigm. The story of change in a *Great Transition* is a tale of how the various actors work in synergy and with foresight as collective agents for a new paradigm.

Dimensions of Transition

A *Great Transition* envisions a profound change in the character of civilization in response to planetary challenges. Transitions have happened before at critical moments in history, such as the rise of cities thousands of years ago and the modern era of the last millennium. All components of culture change in the context of a holistic shift in the structure of society and its relation to nature. The transition of the whole social system entrains a set of sub-transitions that transform values and knowledge, demography and social relations, economic and governance institutions, and technology and the environment (Speth, 1992). These dimensions reinforce and amplify one another in an accelerating process of transformation.

Values and Knowledge

Prevailing values set the criteria for what is considered good, true and beautiful. They delineate what people want and how they want to live. Values are culturally conditioned, reflecting the social consensus on what is considered normal or desirable. Depending on its dominant values, a society lies along a continuum between antagonism and tolerance, individualism and solidarity, and materialism and a concern for deeper meaning. Individualism and consumerism drive the unsustainable trends of *Conventional Worlds*. But they are neither inherent nor inevitable. The plausibility of a *Great Transition* rests with the possibility that an alternative suite of values emerges to underpin global development.

The distinction between "needs" and "wants" has profound implications for the transition. Physiological, psychological and social needs are universal, but culture shapes how they are perceived and how they are expressed as wants (Maslow, 1954). Advertising and media can stimulate new wants and the experience of them as felt needs. Values mediate how needs are transformed into wants and how they are satisfied. The need for sustenance can be satisfied by steak or vegetables. The need for self-esteem can be satisfied by a luxury car or a circle of friends. A value transition to post-consumerism, social solidarity and ecology would alter wants, ways of life and behaviors.

A complex set of factors drives the search for new values. Both angst and desire—the concern about the future and its lure—play roles. Anxiety over ecological and social crises leads people to challenge received values. This is the "push" of necessity (Table 3). At the same time, visions of a more harmonious world and richer lives attract people toward the new paradigm—the "pull" of desire. Together they lead to a revised notion of wealth that underscores fulfillment, solidarity and sustainability.

Individualism, consumerism and accumulation may help the market reach its full potential. But as dominant values in the Planetary Phase, they are shackles on the possibility of humanity reaching its full potential. On the path to a *Great Transition*, awareness of the connectedness of human beings to one another, to the wider

Table 3. Pushes and Pulls Toward a New Paradigm

Pushes	Pulls
Anxiety about the future	Promise of security and solidarity
Concern that policy adjustments are insufficient to avoid crises	Ethics of taking responsibility for others, nature and the future
Fear of loss of freedom and choice	Participation in community, political and cultural life
Alienation from dominant culture	Pursuit of meaning and purpose
Stressful lifestyles	Time for personal endeavors and stronger connection to nature

community of life and to the future is the conceptual framework for a new ethic (ECI, 2000). Taking responsibility for the well-being of others, nature and future generations is the basis for action.

The knowledge transition would expand the ways in which problems are defined and solved. The fundamental units of analysis of a new sustainability science are socio-ecological systems, as they form and interact from the community to planetary levels. These are complex and non-linear systems with long time lags between actions and their consequences. A systemic framework is required to illuminate key problems such as the vulnerability of systems to abrupt change and interactions across spatial scales. Sustainability research defines a fascinating new program of scientific research. It also is the basis for an early warning system that can alert decision-makers and the public on future perils and provide guidance on ways to respond.

The linkages between human and biophysical systems require the unification of knowledge. The reduction of whole systems to their constituent components was an important methodological advance of the scientific revolution. The division into separate disciplines of inquiry was essential for focus and rigor. These are necessary for addressing the complex problems of transition—but they are insufficient. An interdisciplinary focus on holistic models must now complement the reductionist program.

The challenge is to develop appropriate methodologies, train a new cadre of sustainability professionals and build institutional capacity. A science of sustainability would highlight integration, uncertainty and the normative content of socio-ecological problems (Kates et al., 2001). Sustainability science proceeds along parallel lines of analysis, action, participation, policy and monitoring in an adaptive real-world experiment. To be trustworthy, knowledge must be rooted in scientific rigor. To be trusted, it must reflect social understanding. The peculiar nature of sustainability problems requires that diverse perspectives and goals be brought to the scientific process. This requires the cooperation of scientists and stakeholders, the incorporation of relevant traditional knowledge, and the free diffusion of information.

For all this to happen, research and educational institutions will need to encourage, support and professionally reward this type of research. The institutional basis for a knowledge transition must be constructed, especially in developing countries. In this regard, information technology offers unprecedented opportunity to provide universal access to data systems, analytic tools and scientific findings. Scientists, policy makers and citizens can interact through networks of research and exchange. The democratization of knowledge would empower people and organizations everywhere to participate constructively in the coming debate on development, environment and the future.

Demography and Social Change

People, their settlements and their social relationships are undergoing rapid and profound change. Growing populations, expanding cities, a continuing rights revolution and globalization are critical demographic and social trends. These will play out differently in the various scenarios of global development. A demographic and social transition is a critical aspect of the wider enterprise of a *Great Transition*.

Cresting populations and reinventing cities Population growth is slowing. The world population of over 6 billion people is growing at an annual rate of 1.3 percent, adding about 80 million people each year. The peak growth rate of about 2.2 percent occurred in the early 1960s and the peak population increase of about 87 million per year occurred in the late 1980s. Under mid-range assumptions for changes in birth and death rates, population is projected to be over nine billion in 2050 (UNPD, 2001), with almost all of the increase in developing countries.

The acceleration of population stabilization is both an end for, and means to, a *Great Transition*. As an end, decreased birth and death rates can enhance life quality—for children, increased survival, growth and development; for their mothers, lower mortality and greater opportunity for education, work and income; for their fathers, healthier living; and for their grandparents, a longer life. At the same time, the family, the oldest of institutions, is challenged to redefine itself, as siblings diminish and parents age. As a means, a *Great Transition* becomes more feasible in a lower population world. Fewer people would reduce pressure on the environment and reduce the ranks of the impoverished.

The value and social policy changes on the path to a *Great Transition* could decrease projected populations by a billion people by 2050. This would result from satisfying the unmet need for contraception and from parents opting for smaller families and postponing parenting. A key is to join reproductive health services in developing countries with education, particularly for girls, and job opportunities for using their training.

The number of city dwellers has grown much faster than population, with over half the population now urban. If trends continue, the urban share of population could grow to as much as 75 percent by 2050, swelling cities by nearly four billion people, or the equivalent of 400 cities the size of Buenos Aires, Delhi, or Osaka. On average, people who live in urban areas receive more income, have fewer children, have better access to education and live longer than their rural counterparts. But cities are also places of extreme contrast in wealth and opportunity. For the poor of many cities, urban life is more difficult and less healthy than life in the countryside.

The challenge that faces the planners, designers, builders and financiers of expanding cities is also an opportunity. The urban transition is about creating urban settlements that make efficient use of land and infrastructure, and require less material and energy, while providing decent living conditions. The new vision would unify concerns with habitability, efficiency and environment, concerns that are currently fragmented in different agencies and disciplines. Then, the need to replace much of the current infrastructure over the next two generations could become an opportunity to create habitable cities that are resource efficient and ecosystem conserving.

The transition to sustainable urban environments is an immense challenge. The magnitude of the task would be abated to the degree that the demographic transition reduces overall population. Also, a *Great Transition* could diminish urbanization rates by developing more attractive rural alternatives. Communication and information technology would create more flexible options for remote work, reducing the growth of cities. Urban and town settlement patterns that place home, work, commerce and leisure activity in closer proximity would reduce automobile dependence and strengthen communities. The elimination of the urban underclass, and the strengthening of social cohesion would support the transition to diverse, secure and sustainable communities.

Institutionalizing the rights revolution   The last quarter century witnessed remarkable progress toward a consensus on universal rights for people, children, indigenous cultures and nature. These rights protect civilians caught in civil and international conflict, prohibit genocide and torture, forbid hunger as an instrument of war or repression, provide refuge for abused women, proscribe child exploitation, protect endangered species and affirm diversity in both nature and society.

Rights are expressed through international agreements and administered through new institutions. But their enforcement is far from complete. The transition envisions the acceleration of that process—the institutionalization of inviolable rights of people and of nature. One task is to build popular awareness of established rights and to enforce those rights. Another is to expand them through the extension of freedom and democracy.

But rights are often in conflict. The challenge is to respect minority rights, while avoiding fragmentation into separate identities, territories or even species. Armed conflict will not be reduced unless alternative ways of providing ethnic or religious autonomy without fragmentation are developed and widely accepted. Life-support systems will not be preserved without well-recognized rights for nature that go beyond single favored species preservation to embrace natural communities and ecosystems.

The eradication of cruelty toward humans will not be served by condoning cruelty toward animals. The *Great Transition* is a human event and humans are at its center. Meeting human needs or extending life quality inevitably involves finding the balance between the use of domesticated and experimental animals and the universal rights of sentient creatures. Rights in conflict are among the most difficult but meaningful challenges. Over time, humankind learns how to extend rights, resolve some conflicts and live in peace with the remainder.

Poverty and equity   Currently the global economy has a dual character. A dynamic, modern, formal component coexists with a rural, informal livelihood economy. The income of the richest 1 percent of the world's people equals that of the poorest 57 percent, while nearly three billion people live on less than $2 per day (UNDP, 2001). Globalization would threaten further marginalization if local economies are subject to the imperatives of global markets with little commitment to place or people. In that kind of globalization, the egalitarian and democratic aspirations of the modern era would remain unfulfilled.

The social transition would focus on the well-being of the poor, sustainable livelihoods and greater equity. The foundation for a *Great Transition* is a world where human deprivation is vanishing and extremes of wealth are moderating. Then the promise of the twentieth century for universal access to freedom, respect and decent lives may be fulfilled in the twenty-first. As new values and priorities reduce the schism between the included and excluded, the space opens for solidarity and peace to flourish. Poverty reduction and greater equity would feed back to amplify the process of transition.

Economy and Governance

A *Great Transition* implies a revision in human institutions—the relationships and patterns that organize the behavior of a society.

Institutional change would both drive and respond to parallel evolution in values, knowledge and ways of life. Critical to this process would be the changing character of the economy and governance.

Contours of a new economy   The economic transition means moving towards a system of production, distribution and decision-making that is harmonized with equity, sustainability and human fulfillment. It would balance multiple objectives: eradicating human deprivation, reducing inequality, staying within environmental carrying capacity, and maintaining innovation. This would certainly include such policy instruments as eco-taxes, social subsidies and green accounting. But these would be manifestations of deep processes that reorient the way the economy functions. The economy becomes a means of serving people and preserving nature, rather than an end in itself. The transition would be expressed in altered behaviors and practices of people, firms, governments and international governance systems.

As people aspire to sustainable living, purchasing patterns would reflect ecological sensitivity, consumerism would abate and travel patterns would shift toward mass transport. People might increasingly share their time, through voluntary and non-profit work, and their income, through voluntary donations and support for redistribution through taxation. As affluent countries reduce their environmental footprint, resources would be freed for others.

The changes in consumption patterns would send powerful market signals. The self-interest of business remains an important economic engine, but business interests, too, change. Enlightened businesses would increasingly seize the initiative, showing that eco-efficiency, green marketing and social responsibility offer a competitive advantage. Corporations that pursue new codes of conduct would be rewarded in the market place, while those that do not would be punished by an increasingly informed and vigilant public mobilized by NGOs.

In the course of transition, business would gradually revise the bottom-line to include social equity and environmental sustainability, not only as means to profit, but also as ends. Big corporations would play a leading role in this transformation as their huge technical and financial resources provide space for strategic innovation. But with their human links and local roots, small businesses also would be important players.

While substantial investment in environmental and social goals would be required, the world economy has the resources for such an undertaking. Moreover, the transition would mobilize "new dividends." A *green dividend* would flow from the cost-savings of eco-efficient corporations and the maintenance of society's environmental capital. A *peace dividend* would stem from gradual reduction of the world's $700 billion annual military expenditure to a sufficient level for world peace-keeping, perhaps $30 billion (Renner, 1994). A *human capital dividend* would come from harvesting the creativity and contributions of the billions who would otherwise be consigned to poverty. A *technological dividend* would derive from new opportunities for innovation and wider access to the information revolution. A *solidarity dividend* arises from reduced security and police costs.

The economic transition is a matter of will, not resources. If values and priorities were to change, economic resources are at hand.

New institutions   The governance transition is about building institutions to advance the new sustainability paradigm through partnerships between diverse stakeholders and polities at local, national and global levels. While specific structures will remain a matter of adaptation and debate, a proliferation of new forms of participation can be expected to complement and challenge the traditional governmental system. In the new paradigm, the state is embedded in civil society and the nation is embedded in planetary society. The market is a social institution to be harnessed by society for ecology and equity, not simply wealth generation. The individual is the locus of a web of social relationships, not simply an atom of pain and pleasure.

Expansion of individual or household entitlements would address social equity. For example, a minimum basic income could be guaranteed to all, possibly through the mechanism of a negative income tax. This would both reduce poverty and advance gender equality by increasing the economic independence of women. A guaranteed income would indirectly benefit the environment, as well, by reducing the incentive to combat unemployment and poverty through greater economic growth (Van Parijs, 2000). At the other end of the income distribution, progressive taxation would limit individual income and wealth to what *Great Transition* societies find acceptable based on equity and sustainability considerations.

Market regulations would ensure that market forces do not violate social and environmental goals. They would rely on self-regulation by socially and environmentally conscious producers, public pressure, and local, regional, national and international agreements. An empowered and information-savvy network of NGOs, issue-based associations and green producers would reduce the need for government regulation and enforcement.

Income transfers from urban to rural populations could pay for the nature conservation services that the latter perform, such as the European Union policy of paying farmers to maintain rural landscapes. In the transition, analogous mechanisms could transfer resources from rich cities to poor rural areas to simultaneously reduce poverty and secure the provision of ecosystem services such as biodiversity, forest and water conservation, and carbon sequestration.

New roles would evolve for national government in response to pressures coming from all directions. From below, responsibilities would move to local levels in the spirit of subsidiarity and particpation. From above, the expanding needs for global governance would move greater decision-making to the international context. From the side, businesses and civil society would become more active partners in governance. That said, national governments would retain considerable authority, not the least of which would be playing a central role in brokering societal agreements. They would need to do so in ways that are transparent, accountable and democratic.

International negotiation and regulation would grow in importance, since economic, environmental and social issues are increasingly of a global character. These enlarged processes would set and enforce minimum sustainability standards such as basic human entitlements, environmental resource protection and human rights. The strategies for implementing such standards would be left to national and sub-national deliberations, and would take diverse forms depending on political cultures. In addition to formal governmental processes, international discussions and agreements would involve business groups, consumers' associations and other global networks.

The global information revolution would spawn new international experiments to reshape corporate governance. This would include collaborative processes of corporations, governments, NGOs and grassroots organizations. Current initiatives offer a glimpse of new approaches for increasing transparency and accountability, and aligning business practices with sustainability principles that are appropriate to a planetary society.

Key elements for reducing poverty would be egalitarian policies for wealth redistribution and targeted social expenditures for the poor. In addition to macro-policies, civil society programs would work from the bottom up to address poverty from the perspective of the poor themselves. The goal would be to enhance the individual and collective capacity of the poor to cope with their situation. They would channel resources back to the livelihood economy through collective institutions, financial systems and appropriate technology, and would foster cooperation among businesses, NGOs and communities.

Technology and the Environment

The technology transition would sharply reduce the human footprint on nature. The three pillars are efficient use, renewable resources and industrial ecology. Efficient use means radically reducing the required resource inputs for each unit of production and consumption. Renewable resources means living off nature's flows while maintaining its capital stocks—solar-based energy rather than fossil fuels, sustainable farming rather than land degradation and preserving ecosystems rather than liquidating them. Industrial ecology means largely eliminating waste through re-cycling, re-use, re-manufacturing and product life extension. We consider several key sectors.

Energy The challenge is to provide affordable and reliable energy services without compromising sustainability. This is a both a social and environmental transition. The social energy transition would give access to modern fuels to the global billions who still rely on dwindling traditional biomass sources. The environmental energy transition would cut the demand side through moderated consumption in affluent areas, high end-use efficiency and deployment of renewable sources.

The imperative to reduce global greenhouse gas emissions sets the magnitude and pace for the energy agenda. Climate stabilization at safe levels requires transcending the age of fossil fuels. The path to a solar future would be bridged by greater reliance on natural gas, a relatively low-polluting fossil fuel and modern biomass technologies. Nuclear power is climate-friendly, but other problems—long-term radioactive disposal, uncertain safety and links to weapons proliferation—are incompatible with a resilient and sustainable energy future.

The challenge is immense, but so are the technological possibilities. On the demand side of the energy equation, appliances, lighting, buildings and vehicles can be made highly efficient. Combined heat and power systems can capture energy that would otherwise be wasted. Compact settlements can reduce travel and encourage energy-sparing modes of travel, such as mass transport and cycling. The Internet has the potential to substitute information for energy and materials through e-commerce.

On the supply side, solar energy can be captured in diverse forms—directly by solar cells and heating systems, and indirectly through wind, moving water and biomass. Solar energy can be used to generate hydrogen, a clean liquid fuel that can substitute for petroleum in vehicles. Solar technologies have been gradually expanding their market share, but at a snail's pace. They tend to cost more than fossil fuels, but the gap is gradually decreasing, and would close entirely if environmental costs were factored into prices.

The technological remedies are either at hand or could be matured through a re-prioritization of research, development and deployment efforts. Institutional barriers are more serious.

Technological and infrastructural inertia maintains dominant patterns that have locked-in over the decades. Powerful vested interests seek to preserve dominance in conventional energy markets. Perverse policies subsidize fossil fuels and inefficiency. The incentives for a new energy era must be built into policies, prices and practices that can counteract the recalcitrant fossil fuel economy in high-income countries, and allow developing countries to leapfrog to the solar era. In a *Great Transition*, this would become a popular imperative.

Food and land   The goal is to provide sufficient food for all, while preserving soil quality, protecting biodiversity and preserving ecosystems. The "green revolution" had great success in raising crop yields. But heavy use of chemicals has polluted soils and groundwater, and nearly a billion people remain undernourished. Forests and other ecosystems continue to be lost to agricultural expansion as growing populations, higher incomes and more meat in diets require more farmland and pastures.

The agriculture transition would promote farming practices that are more knowledge-intensive and less chemical-intensive. Complex farming systems would build on natural synergies such as nitrogen-fixing plants grown in combination with other crops to reduce fertilizer requirements, and integrated pest management to reduce pesticide use. Soil conservation would maintain quality through efficient drainage, terrace agriculture, conservation tillage and other techniques. Fish farming would adopt strong environmental standards, while marine harvesting would be kept within the carrying capacity of wild fisheries. On the demand side, moderated food demands would lower pressure in a *Great Transition* as populations stabilize and diets shift away from meat.

Biotechnology holds the promise of increasing yields, reducing chemical input, conserving water and improving the nutritional content of crops. But it carries the risks of reducing crop diversity, of degrading ecosystems through accidental release of pest-resistant organisms into the wild, and of increasing the dependency of farmers on transnational agribusinesses. The transition would be guided by the precautionary principle, deploying biotechnology where it can enhance agriculture production in an environmentally sound and safe manner. At the same time, complementary advances would be sought in areas with lower risk and greater public acceptance, such as improved breeding.

The preservation and restoration of the world's ecosystems is a central theme of the transition. This would be supported by the valorization of what ecosystems provide in goods, services, aesthetics and habitat. Trends in land use would change, including controlling urban sprawl through more compact and integrated settlements patterns. The reduced stress from pollution, climate change and excessive water extraction would help maintain the resilience of ecosystems.

Water   Freshwater sustainability seeks to provide sufficient water for human needs, economic activity and nature. Diverse solutions matched to local conditions will be needed to manage demand growth and enhance supplies.

Water requirements can be reduced through efficiency improvements in irrigation and other water using activities; by reducing transmission losses; and by considering non-hydropower generation supply. New crop varieties and improved cropping methods would increase the "crop per drop" for both irrigated and rain fed agriculture. Some arid places would need to rely on greater food imports to reduce local water requirements for agriculture. In areas of high water stress, the lower populations and revised consumption patterns of a *Great Transition* are critical to the water transition.

Intact ecosystems would help to maintain resources by moderating flood runoff and enhancing groundwater storage. Also, greater deployment of unconventional supply methods, such as small-scale water harvesting schemes, rainwater capture, desalination in coastal cities and recycling treated wastewater for agriculture, would contribute. The key is to place the freshwater issue in a systemic framework that comprehensively considers ecological and human needs. Decision-making would move from centralized agencies to the watershed where the allocation of water can best be resolved. The principle of the active participation of stakeholders representing diverse interests would be critical for balanced and resilient solutions.

Environmental risk and development   In the mature industrial countries, the goal is to phase in technologies, practices and infrastructure as capital stock turns over. In developing countries, the goal is to leap to advanced eco-efficient technologies that are well suited to their social and ecological endowments, thus avoiding a recapitulation of the resource-intensive stages of industrialization. Alternative pathways are illustrated in Figure 10. By adopting innovative technologies and practices, developing countries could tunnel below the safe limit.

On the sustainability path, technologies and practices can act synergistically—ecosystem protection sequesters carbon, water conservation reduces soil degradation, renewable energy mitigates both climate change and air pollution. The mandate for applied science and human ingenuity is to radically reduce the flow of materials into the global economy, and the waste that is generated. The available pool of technologies and creative capacity provides a strong platform for launching a technology and environmental transition.

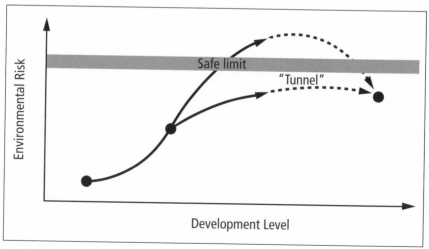

Based on Munasinghe (1999).

Civilizing Globalization

Globalization is more than economic integration. Words, images and ideas outpace the flow of products, raising fears of loss of language, culture and values. Counter-currents emphasize ethnic, national and religious distinctiveness. Flows of people—temporary, permanent and forced—also swell. Refugees and world trade grow at similar rates. The influx of immigrants makes places of wealth and opportunity more diverse, and that has often not been welcome. Diseases move with people and products, affecting human health, crops and livestock. Biological invasions can destroy native biota. Environmental harms are exported to countries with weaker protection. Terrorism becomes globalized. Aggressive marketing and rapid cultural change fuel global consumption. Yet, billions are excluded from prosperity.

But as communication carries a culture of consumption, it also carries a culture of concern with the fate of the earth and future generations. It links people and groups in an expanding project to share information and influence development. The widening, deepening and accelerating interconnectedness that characterizes globalization is the precondition for a *Great Transition*. Globalization forges expanded categories of consciousness—seeing humanity as a whole, its place in the web of life and its links to the destiny of the planet. It distributes systems of production and participation, creates potential roles for corporate and civil society and makes greater equity possible.

For those who aspire to a more humane, sustainable and desirable future, simply being "against globalization" is not satisfactory. Rather, the struggle is over the character of globalization in the coming decades. If its promise is to be realized and its perils avoided, globalization must be reshaped. A *Great Transition* needs globalization and needs to deal with its discontents. The victims of globalization and their many concerned allies are doing more than demonstrating in the streets. They are also developing an understanding of what is needed to civilize it (Held et al., 1999; Helleiner, 2000).

The principles and means for shaping a new kind of globalization are in place. A *Great Transition* would find rights assured, nature treasured, culture rich and the human spirit animate.

5. History of the Future

Dateline: Mandela City, 2068

A century ago, the Apollo 8 mission first transmitted the image of our Blue Planet, a beautiful and delicate pearl floating amidst the dark cosmos. This icon from space gave vivid witness to the fragility and preciousness of our common home, and was forever riveted on the human imagination. But it could not reveal the great changes that were quietly building, changes that were destined to transform human history and the Earth itself.

Prologue

From the current vantage point, with the planetary transition unfolding before our eyes, it would be premature and vain to attempt a definitive account of this extraordinary era. Our history remains the subject of energetic debate among twenty-first century scholars, complexity specialists and a public whose fascination with the past knows no bounds. But the past remains ambiguous, while the future defies prediction—who can say what new surprises await us? The task of analyzing the causes and significance of our tumultuous century must be left to future historians, who can tell the story with greater objectivity, subtlety and wisdom. In this brief treatise, we can offer only a thumbnail sketch of the broad historical contours of what we have come to call the *Great Transition*, and our admittedly subjective observations on the momentous events that shaped it.

With a long view, our century of transition is but a moment in a long process of human evolution. We think of earlier great transitions—Stone Age culture, Early Civilization and the Modern Era—as fulcrums in time when the very basis of society was transformed. To this august list of celebrated milestones along the path of human history, we may now add, in our judgment, a new one. The planetary transition has ushered in a new stage of social complexity, culture and novelty. For the first time, the dynamics of human development must be understood as a phenomenon occurring at the global scale. Where earlier transitions evolved slowly over many millennia or centuries, this one occurred in a heartbeat of historic time. Where change once radiated from local innovation, this was a transformation of the global system as a whole, involving all the world's peoples and, indeed, the whole community of life on the planet.

The immediate antecedent for the *Great Transition* was the industrial revolution. Centuries of institutional, cultural and technological change during the Modern Era prepared the groundwork. Then, the industrial explosion launched an exponential spiral of innovation, economic expansion and population growth, the Big Bang that propelled humanity toward its Planetary Phase. As industrial society inexorably grew, it absorbed traditional societies on its periphery into the market nexus and pushed against the boundaries of the planet's environmental capacity.

Wherever it went, industrial capitalism left a contradictory legacy. In part, its story is an emancipatory tale of wealth generation, modernization and democracy. But it's also a heartless saga of social disruption, crushing poverty and economic imperialism. Not surprisingly, oppositional movements arose to challenge its human injustices and environmental devastation. Socialists the world over struggled for an egalitarian society where wealth was generated for people instead of profits and where a collectivist ethos replaced the greed of the profit motive. But that dream was dashed by real-world socialist experiments. Challenged militarily and isolated economically, they degenerated into bureaucratic tyrannies that eventually were re-absorbed by the global market system.

In 1948, the signing of the Universal Declaration of Human Rights crystallized the soaring aspirations of a generation. World peace could rise from the agony of world war, the family of man could temper the wounds of hatred, and the bells of freedom could ring in every land. The vision was to be postponed through the long years of human suffering. But it remained a beacon of hope to illuminate the path ahead.

The planetary transition accelerated after 1990 when the fall of the Soviet Union released the world from the stasis of the Cold War. With this major impediment removed, the march of capitalism toward an integrated world system accelerated. Developments in the prior decades set the stage—the birth of the technological underpinnings of the information and communications revolution; the proliferation of international institutions following World War II; the rise of civil society as a "third force" in world affairs; the widespread spiritual revivals and environmental movements that foreshadowed the values-led movements of our own century; the mounting human impacts on the environment that began to trigger planetary-scale processes; and the integration of the global economy catalyzed by increasing international flows of trade, finance and information.

We track the story of transition as it evolved through several phases. The first phase began with the euphoria of market-driven globalization, was punctuated by terror, and ended in despair. The Crisis that followed fundamentally changed the course of global development. Global Reform was a time of renewed attempts at global governance through official channels. Then the *Great Transition* phase ushered in the values-led, bottom-up resurgence of our own time.

Market Euphoria, Interruption and Revival

1990–2015

In the 1990s, an economic growth surge was fueled by the maturing of information and communication technologies into the first flowering of a network economy. The global media were abuzz with a giddy enthusiasm that was difficult to escape. Business gurus, technological forecasters and cultural critics alike pontificated on the new era of "frictionless capitalism." A surging bull-market banished memories of the business cycle. An endless stream of digital gadgetry renewed an orgy of consumption. A globalizing economy was constructing a planetary emporium, bringing Western modernism and dollars to the underdeveloped. A richer world would apply the magic of the market to saving the global environment.

It was never thus. The start-up companies of the dot-com billionaires were dripping with red ink. The then-popular thesis of the "end-of-history" was a comforting ideology for celebrants of capitalist hegemony, rather than serious scholarship. The quest for material excess could not long provide a satisfactory basis for people's lives. Globalization fed new forms of anger and resistance, rather than ameliorating polarization. The market's magic had its powers, but they did not include the foresight and coordination required for environmental sustainability.

In fact, the market euphoria was confined to a small but vocal minority with great access to media, great power to shape public perceptions and great influence on political agendas. Nevertheless, during the 1990s, a loose coalition of environmental, labor and social justice groups held demonstrations against the international economic organizations of that time. The increasingly militant protests challenged a "corporate globalization" that they saw as socially unfair and environmentally insensitive, and threatening to sacrifice hard-won safeguards on the altar of global competition. This early protest movement was fragmented and lacked a clear positive vision of a humane and sustainable alternative. But it was a portent of what was to come. The long struggle over the meaning and character of globalization had begun.

By 2002, the irrational exuberance of the 1990s had vanished as quickly as it arrived. In the first years of the new century, economic retrenchment, bear markets and global terrorism sobered the intoxicated. It had been a "false boom" that was largely confined to the United States, its allies and a few of its supplier countries in Southeast Asia. The economic base was small—less than 5 percent of the world's population had access to digital networks—and the predictable market excesses led to a downturn. The benefits of economic integration were confined to a global elite. At the same time, growing concerns with the environment, persistent global poverty and the culture of consumerism were expanding the popular challenge to the market consensus, especially among youth.

The denouement of naïve market euphoria came in 2001 with the horrific "9/11" terrorist attacks on the citadels of global financial and military might—the World Trade Center and the Pentagon in the United States. This traumatic rip in the culture of complacency awoke the world to the depths of anger fermenting among those exposed to globalization but excluded from it. The desperation of billions was revealed as a fertile seedbed for indoctrination and fanaticism by cynical self-styled Islamic fundamentalists. Where an arrogant West seemed to offer little more than indignity, transnational Islamist organizations could offer the salvation of martyrdom in the armies of global Jihad. Terrorism, too, had gone global.

At the time, it was widely feared that the terrorists would succeed in sparking a spiral of violence as the United States, joined at least passively by virtually all the nations of the world, hit back hard with its War on Terrorism. But instead—and this is the central irony of this period—the international mobilization brought a more mature and realistic form of market-driven globalization. At first, two polar theories on the root cause of terrorism were proffered— too much modernism and not enough. On the one hand, militant fundamentalism, with its violent rejection of tolerance and pluralism, was understood as the dying gasp of traditionalism as it resisted assimilation into the modernist project. As such, it could be exterminated but not palliated. On the other hand, terrorism revealed a great anger on the streets of third-world cities that indicted the failure of modern development, not its success. A globalization that tantalized a global underclass with images of prosperity, but failed to provide opportunity, was surely a recipe for anger and violence.

Correspondingly, the nations of the world, acting in coalition and through the United Nations and other intergovernmental bodies international, adopted a two-prong "carrot and stick" strategy. The "carrot" took the form of major new initiatives to modernize poor countries and bring the moderating influence of market institutions to the masses. The "stick" was the elimination of hard-core fanatics and their organizations through coordinated covert action and, as needed, military assault. Both elements were partially successful. The War on Terrorism gradually destroyed the capacity of global terrorism to mount sustained large-scale attacks. However, sporadic violence, a sense of peril and heightened security became a way of life as the romance of martyrdom drew an endless trickle of alienated youth.

The affirmative program for expanding modern market institutions has been dubbed the Era of Inclusive Growth. Between 2002 and the Crisis of 2015, a redoubled international effort to promote trade liberalization, modernization and extension of market institutions launched a new wave of globalization. Chastened and more modest in ambition, the second wave brought economic growth almost everywhere and gradually installed a new generation of modernizing technocrats in most countries of the world. The set of policy strategies was not new—the International Monetary Fund had been promoting structural adjustment for years, and the WTO had advanced open markets. But the sense of urgency and level of resources was unprecedented. Before 2002, the United States and some of its allies had been drifting toward an unstable mix of economic globalization and political isolationism. After 2002, they had re-engaged in a vast project to build an interconnected and law-governed global market system.

Debt was forgiven on a strategic basis, new flows of foreign assistance supported modernizing forces in the most underdeveloped countries, nation-building initiatives created more stable regimes and peacekeeping forces maintained stability. By the time of the Crisis, networks were spreading far and wide, and user-friendly technologies like voice recognition and touch screens with universal graphic interfaces had extended at least some access to nearly half of the earth's then seven billion inhabitants. The wiring of the world is a justly celebrated achievement—ironically the "wires" were optical fiber and wireless links. Although the network did not become truly universal until later, the economic upsurge of the period was dependent in part on the global extension of the digital infrastructure.

It was a time of powerful corporate giants whose reach spanned the globe and who could increasingly out-maneuver and influence national governments. The digital giants that built the infrastructure and wrote the software, the consumer product companies that used those links as distribution channels to reach ever-larger markets, the energy behemoths that fueled and powered the boom and transported its products, the global banking and securities firms that financed the expansion—all generated enormous wealth and reached a size and power unprecedented before or since.

A number of important global governance initiatives paved the way. The WTO provided the legal basis for the global trading system. A multilateral agreement liberalized investment regimes, first in the rich countries and then throughout the world. Barriers to trade and capital movements gradually vanished as a host of international instruments promoted market openness and global competition. Almost all national governments were able to overcome internal resistance to aligning their institutions with the imperatives of globalization. They steadily advanced a policy package of modernization of financial systems, public education reform for the new global economy and privatization.

But beyond promoting economic globalization and keeping the peace, global governance became increasingly irrelevant. Of course, international negotiation continued on critical environmental and social problems. But they were either vastly insufficient like the Kyoto Protocol on greenhouse gasses, or nothing more than rhetorical appeals for "sustainable development" and poverty reduction, with little programmatic and financial follow-through. The ideology of "inclusive growth" was compatible with efforts to build the enabling institutions for market progress, but not with proactive pursuit of such non-market goals as environmental sustainability and poverty reduction. Faith in market solutions and trickle-down economics—backed by security and military apparatus—prevailed among powerful world institutions and leaders.

The world became increasingly more integrated culturally as well as economically. The values of consumerism, materialism and possessive individualism spread rapidly, reinforced by communications media. In some countries, fears of being engulfed by Western culture ("McWorld" was the pejorative of choice) continued to stimulate strong traditionalist reactions. But, except for notable fundamentalist strongholds, the lure of the God of Mammon and the Almighty dollar proved too strong, especially while the boom continued and prosperity spread. The protest movement against corporate-driven globalization continued and even grew. But failing to put forth a credible positive vision and strategy for development, it could not galvanize the mainstream in rich or developing countries, and lost political traction.

Although huge inequities persisted throughout this era, economic and digital globalization brought benefits to many, often in unforeseen ways. For example, virtual banking originated in the rich world as a convenience to customers, but in the developing world made a whole suite of financial services available in poor communities where there were no banks or other sources of credit. With the proliferation of digital networks, on-line micro-finance organizations grew rapidly. With credit and connectivity—and decreased corruption due to more transparent virtual banking—came an explosion of small-scale enterprise and increases in productivity. Digitally-connected farmers learned about improved techniques, got loans to buy more productive seeds, used weather information to guide planting and harvesting, and checked market prices for their crops before deciding when and where to sell. Artisan cooperatives could sell traditional handicrafts or made-to-order clothing to major retailers and customers a country or a continent away. Small manufacturers, merchants and service providers expanded to become regional competitors.

Real incomes, even in some poor communities, rose rapidly, radiating out from those countries, such as India, China, Brazil and South Africa, that early on had embraced universal digital access and open commerce. The gradual convergence of the developing world toward the standards of rich countries—the Holy Grail of conventional development thinking—seemed a plausible though distant possibility. But the laser-beam focus on economic growth had dark sides, as well. The market jubilation emanating from the media and the public relations machines of multinational corporations drowned out the voices of concern. All the while, the signals of ecological instability, biological destruction and human health risks became stronger and more frequent. Mounting environmental changes—a warmer and more variable climate, collapsing ecosystems, failing fisheries—hurt poor communities most. Scientists warned with increasing urgency that stresses on the global environment could be approaching thresholds beyond which catastrophic events could ensue.

Billions of the global poor left behind by the boom were growing restive. As the rich got richer and new social strata achieved affluence, deep poverty still chained billions to meager existences. Income distribution became more unequal. Nearly a billion people still went hungry, a figure that put the lie to the homily of market ideologues that the "tide of economic growth would lift all boats." With greater connectivity, the growing disparity between rich and poor was increasingly visible to both. As migration pressure, anger and political dissent mounted, wider social unrest and conflict seemed to loom. Infused with new support, anti-globalization groups increased their agitation for a new direction of social and environmental renewal.

The Crisis

2015

Eventually, as all booms do, the period of market-driven growth came to an end. With the benefit of hindsight, the Crisis of 2015 might seem like a predictable consequence of the tensions and contradictions that had been brewing in the preceding decades. But life is lived forward, not backward, and what seems inevitable in retrospect, in fact, took the world by surprise. The reforms of the Era of Inclusive Growth had their successes—modern institutions and economic expansion were extended to most countries and terrorism was managed at tolerable levels. But they failed to address deep crises that were maturing in proportion to the success of the global market program. Environmental degradation, social polarization and economic distortions were on a collision course, but in the midst of market frenzy, few were able to see it coming.

The Crisis had multiple causes. The bite-back from resource degradation and ecological disruption imposed growing costs on people, ecosystems and the global economy. The collapse of major fisheries contributed to food shortages and stressed international food programs; water shortages grew acute in many places, requiring costly efforts to maintain minimum standards; and resource costs, such as forest products for paper and packaging, rose sharply. While elites emerged even in the poorest countries, persistent poverty and social polarization were eroding the very basis for rule-governed market-driven development. As disparities became more extreme and more visible, social protest and even violent riots became widespread, the march of a million displaced fishermen on New Delhi and the water riots in Iraq being notable examples. Aggravated by environmental and social crises, and with little global governance capacity to respond beyond traditional and ineffective monetary and fiscal measures, the expected contraction after the long global boom triggered a general economic crisis.

The Crisis unleashed a widespread social revolt against the dominance of global corporations, against a quarter century of appalling environmental degradation, and against the persistence of poverty and social squalor amidst great wealth. The Crisis released all the discontent and apprehension about the drift of global development that had been building beneath the surface since the 1990s. The consensus underpinning the era of Market Euphoria was rapidly unraveling. Especially for the world's youth, it was a revolt against what they saw as the soulless materialism and inequity of the established global order. It was at this time that the Yin-Yang Movement was formed out of separate cultural and political youth movements (see box below). Although it was derisively referred to as the Children's Crusade at the time, the unified youth movement was a critical partner in the coalition for a new global deal that led to the Global Reform process.

Under the category "what could have been," it is worth mentioning here the abortive movement known as the Alliance for Global Salvation that arose at this time. The Alliance included a motley group of global actors from the corporate world, the security community and right-wing political elements. Concerned that the crisis could spiral out of control, they came to the conclusion, many reluctantly, that the vacuum of international control must be filled, and that they were the ones for the job. Ironically, this authoritarian threat served to further galvanize the reform movement that warned against the danger of a *Fortress World* "solution." A century before, a previous experiment in globalization had collapsed into the tragedy of the Great War. The forces for a democratic renewal were determined to thwart another return to barbarism.

Global Reform

2015–2025

One of the indirect impacts of the global boom was the expansion and consolidation of democratic governance at national and local levels. Information and communications technologies gradually improved the efficiency of government, allowing people to vote; pay taxes; register land, vehicles, births and deaths; and file complaints more readily and in more transparent ways. Pressure from a more informed and prosperous citizenry—and often from global companies—became harder to resist. Both demanded more responsive governance and more reliable enforcement of laws. Those with holdout dictators or repressive regimes became increasingly isolated.

By 2015, governments were ready to assert themselves on behalf of their citizens. As political leaders everywhere sought to cope with the Crisis, the result was an eruption of governmental leadership at national and local levels. The response took many forms, as governments found ways to re-establish order, to rein in the giant corporations, to clean up the environment, to improve equity, and address persistent poverty and a host of other concerns. This burst of governmental leadership was echoed on the international level.

Before the Crisis, global governance was effective primarily in one area—setting the terms for liberalized trade, de-regulation and privatization. But the renaissance that occurred during the Global Reform era went far beyond anything known before. The World Court, the reconstituted World Union (formerly the United Nations) and the World Regulatory Authority (descended from the last century's Bretton Woods institutions)—all date from this period.

As the world struggled to regain its economic footing, while altering the rules for economic activity, these strengthened institutions offered a new basis for regulating the global market. Chastened by the crisis and buoyed by the popular outcry for leadership, world leaders acted decisively. Sustainable development, the half-forgotten battle cry of the late twentieth century, was resurrected. But instead of rhetoric, a comprehensive set of environmental and social goals were set and the policy muscle was enacted to enforce them.

Treaties were negotiated on global caps and trading regimes for climate-altering emissions, strict limits on ocean fisheries, and outright bans on international trade in wood and other products from endangered ecosystems. Small taxes were imposed on trade and international currency flows, and the revenues used by the world's governments to fund international health, education and environmental restoration. Innovative and generous programs to reduce poverty and provide sustainable livelihoods to all were launched. In a landmark ruling, the World Court asserted jurisdiction over an antitrust case against the world's largest energy company, and subsequently ordered it broken up into half a dozen separate companies, setting a precedent that was applied in many areas of commerce.

By 2020, global economic growth had resumed, not in spite of the imposition of sustainability goals, but because of it. Orchestrated by the new governance institutions, the massive projects to complete the unfinished business of Wiring the World, investing in the poor and saving the environment proved to be a stimulant to an unprecedented period of economic expansion and technological innovation. But this new boom was different from its predecessor. Instead of disparities between North and South increasing, the gap was closing through global programs targeted at raising the standard of living of the poor. Instead of national income distributions becoming more unequal, the gap between rich and poor within countries was either maintained or gradually decreased. Instead of environmental heedlessness, under activist governments the pressure on natural resources and ecological systems began to abate.

The age of sustainable development had arrived, but not for long. Although it created institutions and reforms that have continued to play an important role, the era of Global Reform was relatively brief. Its golden years were from 2015–2020, when the need for post-Crisis recovery led to the strong political unity needed to maintain the reform process. Multinational corporations, seeing their markets stagnate, got on board. But once the boom resumed, many business leaders advocated a return to free markets and a weakening of reforms. By contrast, environmentalists, pleased at first with the accomplishments of the reform agenda, eventually came to see the imposition of restraint on the global growth machine as inadequate—like going down the up escalator. The inherent political and environmental tensions of the forced marriage between sustainability and market growth deepened.

Governments could not keep up with the complex and rapidly shifting concerns of their populations. The public confidence in top-heavy government management mechanisms eroded as the limits of government-led Global Reform institutions to effectively address the complex task of global sustainability became clear. A new cause would appear out of one on-line discussion forum or another, sweep across cyberspace and the media, dominate political discussion with demands for immediate action—and then, just as suddenly, disappear while government was still struggling to act. And global governance through the formal international institutions proved inadequate to monitor and influence rapidly changing social and industrial practices across the kaleidoscope of two hundred nations. More fundamentally, as mathematicians have since established, deterministic management of an often chaotic, non-deterministic, multiplayer system is simply impossible. Policy reform made a difference, as did strong and competent governments, but neither proved adequate to make the changes increasingly demanded by the world's peoples.

At a global scale, building the consensus for new treaties, or even for allocating global funds generated from existing mechanisms, became ever more difficult and contentious. The bureaucracies that evolved to implement the global regulatory regimes became ever larger and more ponderous. A number of countries simply opted out of some treaties, creating loopholes in enforcement that weakened the new international regimes. Global companies proved very agile in adapting to internationally imposed bans or restrictions, while not fundamentally changing their practices.

The clear lesson of the Market Euphoria era was that footloose market-driven globalization was simply not viable. The government-led post-crisis reorganization restarted economic growth and tamed environmental impacts, while bringing up the bottom of the social pyramid. But by the mid-2020s, Global Reform was losing momentum as the will of political leadership waned, governance became enfeebled and the dream of sustainable development was threatened. Another crisis loomed on the horizon.

A growing global coalition of individuals and organizations came to the conviction that reform was not enough. Fundamental notions were challenged—that endless economic growth could be harmonized with ecology, that consumerism could coexist with a sustainability ethic, and that the pursuit of wealth was the path to the good life. The coalition mushroomed into a planetary mass movement for basic change. Sometimes called the Coalition for a Great Transition, it was more popularly known by the name we use today, "The Bouquet," which of course referred to its icon and its slogan ("let a thousand flowers bloom").

The coalition included civil society in all its stunning diversity—spiritual communities, Yin-Yang, networks of special-interest organizations. All parts of the world community were represented—communities, nations, regions, river basins—in a kind of spontaneous global assembly from below. The basis of their unity was a common set of values—the rights of all people to a decent life, responsibility for the well-being of the wider community of life and the obligations to future generations. The project for more just, more ecological and more fulfilling ways of life was not to be denied.

The Yin-Yang Movement

The youth of the world played a critical role throughout the long transition. Young people have always been the first to take to new ways and to dream new dreams. And so it was with communications technology and the exploration of the possibilities for a new global culture. The main manifestation in the first blush of market euphoria was, of course, the promotion of a consumerist youth culture. But other consequences of the digital information revolution were equally important. The pedagogic impacts of accelerated learning and information access had a great democratizing effect that empowered younger generations to participate fully in the economy and all aspects of society. By 2020, the vast majority of the world's secondary and university students used the Internet as a matter of course, and websites and wireless portals in more than 200 languages catered to them.

The huge surge in Internet-ready young people graduating from schools in the developing world had some unexpected effects. To ease its chronic shortage of skilled workers and take advantage of lower salaries, the burgeoning digital industry increasingly moved its programming, web design, e-learning courseware and other software tasks to India, China and other centers of talent. Leadership of the industry began to follow. And this new leadership played a major role in providing digital services designed for poor communities.

Even more unexpected were the cultural and political changes that universal access set in motion. Internet-powered awareness of a wider world and access to unlimited information accounted for part of the change. Equally important were the proliferation of ways to communicate across cultures and even—with automatic translation—across language barriers through e-mail, mobile phones and messaging networks, and through swapping music, videos, underground political tracts and calls for protest demonstrations in huge informal networks.

The gradual coalescence of a discernable global youth culture is difficult to date. But certainly by 2010, two broad streams had emerged to challenge the prevailing market paradigm. The YIN (Youth International Network) was a cultural movement that advanced alternative lifestyles, liberatory values and non-materialistic paths to fulfillment. The YANG (Youth Action for a New Globalization) was a loose political coalition of activist NGOs that eventually were forged into a more cohesive network through a long series of global protests and actions.

Before 2015, there was some tension between the two strands. To many YANGs, the YINs seemed hedonistic, apolitical and complacent, the heirs to the legacy of 1960s hippies and Timothy Leary. For their part, the YINs saw the YANGs as humorless politicos, who were playing the power game. But the rhetoric of the spokespeople for the two tendencies was more polarized than the participants. In fact, the YIN global celebrations and festivals increasingly had a political tonality. At the same time, the huge YANG demonstrations and protests were as much cultural as political events.

(continued)

The Yin-Yang Movement (continued)

During the Crisis of 2015, these distinctions evaporated entirely. The aspirations that each expressed—the search for more fulfilling lifestyles and the quest for a sustainable and just world—became understood as two aspects of a unitary project for a better future. The Yin-Yang Movement was born.

Many activists saw their movement as a global echo of the youth revolution of the 1960s, an explosion of youth culture, idealism and protest. But in truth, it was far more. The Movement was vastly larger and more diverse than its predecessor, and far more globally connected, organizationally adaptive and politically sophisticated. Without it, what would have emerged from the post-2015 world? Perhaps a descent into chaos; perhaps the authoritarian forces for world order, which were waiting anxiously in the wings, would have triumphed.

While counterfactuals are always speculative, it is certainly clear that in the absence of the Yin-Yangs history would have taken a different turn. The Movement was critical at two key moments in the transition. First it provided a base for the new political leadership that was able to fashion the Global Reform response to the Crisis. Later, throughout the 2020s, it carried forward the spirit of 2015, expressing the new values and activism of civil society, culminating in the landmark changes of 2025, and the consolidation of the *Great Transition*.

Great Transition
2025–

The values-driven movements of our time have their antecedents in the human rights and environmental movements that go back to the twentieth century and the spiritual revivals of this century. The search for meaningful and fulfilling lives and alternatives to materialistic lifestyles has deep historical roots. But only in our era, when the dream of a post-scarcity society that could provide enough for all became a practical possibility, could a post-materialist ethos gain a popular basis.

In the cultural revolution of the mid-2020s, lifestyles and even tastes began to change. For one example, traditional families, now shrunken in size as populations stabilized, and extended in time as populations aged, evolved as values of caring and support extended to more of humankind and even to other species. Or for another, the modern "sustainable diet" movement, which resurrected last century's slogan "you are what you eat," reflected the new vegetarianism. This was reinforced by environmental and health concerns that had given rise to organic agriculture and the animal rights movement.

Increasingly, people took pride in living lives that were rich in time, and sufficient in things. The cultivation of the art of living displaced consumerism as the pathway to happiness and status. The anachronisms of the past, such as immense private vehicles with a thousand gadgets, found a home in museums of cultural history, not in people's lives. The sense that individuals are responsible for what they consume was pervasive.

The values movements touched sympathetic chords throughout the world and were amplified by the discussion forums and rapid global communications on digital networks. The "equal participation" movement that has contributed so much to the openness and accountability of political and institutional processes today drew its inspiration both from anti-poverty activists and from earlier civil rights movements. But sympathy alone does not always translate into action. It was the globalization of civil society—the proliferation of global networks and alliances of Value-Based Organizations (VBOs) dedicated to action—that provided the staying power for permanent change. This was a simple but fundamental transformation in world history—the willingness of people, individually and in groups, to take responsibility for solving problems themselves. This phenomenon has become a defining characteristic of the current era.

Information has always been a source of power, and by 2025 power was shifting rapidly. Global networks of VBOs, armed with digital cameras and other sensors, proved to be the ideal counterforce to predatory global corporations and incompetent governments. They organized vast networks to monitor corporate behavior—how and where they logged forests, the quality of their working conditions and wages, and their contributions to local communities. The information was posted on the Internet, often with video footage. They pressured retailers to shun offending companies and consumers to boycott their products. The VBO networks brought powerful market pressures on global companies. Governments who failed to provide basic services to the poor, to protect environmentally sensitive resources or to uphold universal rights, were the objects of equally powerful political pressures.

By enforcing transparency and demanding accountability, these bottom-up networks of activist citizens provided a rapid and powerful social feedback mechanism, far more potent than formal regulatory efforts of governments and intergovernmental bodies. One global banking firm that denied services to a particular Moslem sect in Indonesia, found thousands of its offices around the world shut down by protestors and its brand name badly damaged. A repressive African regime, targeted by a global ad hoc alliance of VBOs, found itself trying to combat hundreds of web sites with damaging video linked to names, photos, and unflattering bios of the president and senior military officers, as well as the names of global companies who were the primary buyers of the country's products (and who hastened to cancel their contracts).

The accountability movement accelerated a leadership transition already underway in corporations and governments alike. More and more, business leaders not only accepted the legitimacy of many social and environmental demands, but found creative business approaches to meet them. Hundreds of global manufacturing firms adopted "zero impact" goals and met them—producing no waste and releasing no pollution in their worldwide operations, and accepting responsibility for post-consumer recovery and recycling of their products. A number of large firms found ways to cut costs dramatically in order to provide affordable basic goods and services, and often jobs, in poor communities—in the process creating large new markets for themselves. Others employed new nano-technologies to produce better products with far less raw materials and energy; "reindustrialization," as it came to be called, aimed at more sustainable ways to provide the material support to human civilization.

For governments and other official institutions, the accountability movement meant not only far more transparency, but expanded participation in decision-making processes of all kinds. Proposed new regulations or laws were now routinely posted on electronic networks for widespread comment and debate before adoption; so were terms of logging or mining leases on public land or plans to develop natural resources. Elections themselves were mostly electronic, making election fraud far scarcer.

The personal and philosophical dimensions of the *Great Transition* complemented and reinforced these changes. Since the Yin-Yang Movement, the disenchantment of youth with consumerism as an organizing principle for their lives and communities had been spreading. Increasingly, people explored more fulfilling and ethical ways of life that offered a renewed sense of meaning and purpose. In the wealthier areas, the values of simplicity, tranquility and community began to displace those of consumerism, competition and individualism. Many reduced work hours in favor of increased time to pursue study, artistic endeavors, interpersonal relations and craft production. Throughout the world, a cultural renaissance, rooted in pride in, and respect for, tradition, and an appreciation of local human and natural resources, unleashed a new sense of possibility and optimism.

The accountability movement, the widespread sense of individual responsibility, the newfound corporate stewardship on environmental and social issues, the readiness (especially among young people) to protest injustice, the search for culturally rich and materially sufficient lifestyles—all of these marked the emergence of what we now think of as the planetary ethic. While history has not ended, a new foundation for the future has been laid. Poverty still survives in small pockets around the globe, but its eradication is in sight. Conflict and intolerance still flare, but effective tools for negotiation and resolution are in place. Our ailing planet has not yet healed from its environmental wounds, but the world is mobilized to restore it to health. The lure of economic greed and political domination has not vanished, but powerful feedback mechanisms are in place to protect the core commitments that continue to shape our era—the right of all to pursue a high quality of life, cultural pluralism within global unity and humanity as part of a vibrant community of life on planet Earth.

Epilogue

We who live in yesterday's tomorrow can know what those who once speculated on the planetary future could not. Turn-of-the-century prophesies of global calamity have been refuted by choices people made both politically and personally. The exuberance of market optimists, who once wielded such influence, has long ago been revealed as a dangerous absurdity. The utopian dreams of a post-capitalist paradise have also, as they must, been defied. The old reformers, who gathered at Earth Summits and a thousand conferences to design management strategies for a sustainable and humane future, could take us only part of the way. But we are forever grateful for their foresight and commitment, for they gave us, their descendents, the gift of choice.

The timeless drama of life continues, with all the contradictions of the human condition; the hopes and heartbreaks, the triumphs and failures, the beginnings and endings. But the drama unfolds in a theater of historical possibility that few would trade. It is little wonder that we so abundantly honor the struggles and achievements of our parents and grandparents. Now our own generation grows perplexed and troubled by the youth of today, with their cultural rebelliousness, political restlessness and search for new challenges. Are they the harbingers of a new transition struggling to be born? Time will tell.

6. The Shape of Transition

Depending on how the uncertainties of planetary transition are resolved, the global future can branch into distinct paths. The scenarios discussed in this essay are alternative stories of the future, each representing a unique combination of institutions, values, and culture. The narratives can be further elaborated with a quantitative sketch of how key indicators unfold over time. We focus on four of the scenarios—*Market Forces*, *Policy Reform*, *Fortress World* and *Great Transition*.

All scenarios begin with the same set of contemporary trends that are now driving the world system forward. Social, economic and environmental patterns then gradually diverge as they are conditioned by different events, institutional change and value choices. *Market Forces* is a world of accelerating economic globalization, rapid spread of dominant institutions and values, and minimal environmental and social protection—the competitive global market shapes the planetary transition. *Policy Reform* features government initiatives to constrain the economy in order to attain a broad set of social and environmental goals—sustainability policy shapes the planetary transition. *Fortress World* envisions a period of crisis leading to an authoritarian and inequitable future—tyranny shapes the global transition. In *Great Transition*, a connected and engaged global citizenry advances a new development paradigm that emphasizes the quality of life, human solidarity, and a strong ecological sensibility—new values shape the planetary transition.

Global patterns are compared in Figure 11 (Raskin et al., 1998; Kemp-Benedict et al., 2002; PoleStar, 2000). *Market Forces* are defined by counteracting tendencies. Technological innovation steadily reduces the environmental impact per unit of human activity, but the increase in the scale of human activity drives impacts higher. Economies in poor regions grow rapidly, but so do disparities between and within countries. The result is a continued erosion

Figure 11. Scenarios Compared: Selected Indicators

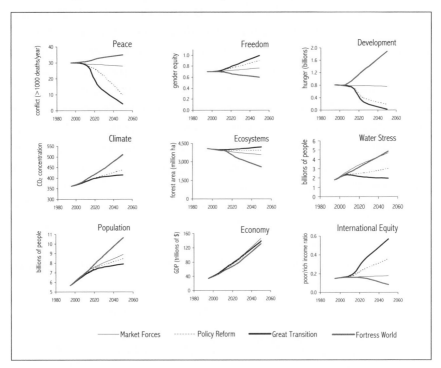

of environmental health and the persistence of poverty. *Policy Reform* "bends the curve" through the rapid deployment of alternative technology—eco-efficient industrial and agricultural practices, highly resource efficient equipment and renewable resources—and targeted programs to reduce poverty. *Fortress World* is a dualistic world of modern enclaves of affluence for the few, and underdeveloped areas of destitution for the many.

Great Transition includes the rapid penetration of environmentally benign technologies, as does *Policy Reform*, but at a more rapid pace. A second major feature also supports environmental sustainability—the shift toward less materially-intensive lifestyles. Resource requirements decrease as consumerism abates, populations stabilize, growth slows in affluent areas, and settlement patterns become more integrated and compact. At the same time, poverty levels drop, as equity between and within countries rapidly improves.

Great Transition patterns are shown in Figure 12 for "rich" and "poor" regions, essentially the OECD countries and the rest-of-the world, respectively. Population growth moderates in response to poverty eradication, universal education and greater gender equality. In affluent regions, income growth slows as people opt for shorter formal workweeks to devote more time—an increasingly valued resource—to cultural, civic and personal pursuits. Rapid investment and transfers to poor regions stimulates rapid growth and international equity. The affluent reduce the fraction of meat in diets for environmental, ethical and health considerations. National equity in most countries approaches the levels currently seen in European countries such as Austria and Denmark. Reliance on automobiles decreases in rich areas, as settlements become more integrated and alternative modes of transportation more prevalent. The energy transition ushers in the age of renewable energy, the materials transition

Figure 12. *Great Transition* Patterns

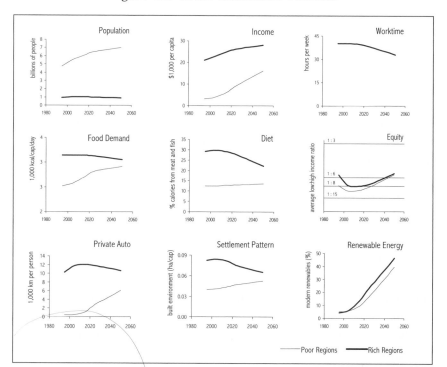

radically reduces resource throughput and phases out toxic materials, and the agricultural transition brings greater reliance on ecological farming.

The *Great Transition* is a complex story. Just as aspects of all scenarios are simultaneously at play today, the world system will unfold as a mixed state as the various tendencies compete for dominance. One possibility for the phased emergence of a *Great Transition* is reflected in the three eras of the "history of the future" (Section 5). The overlay and sequence of scenarios is illustrated in Figure 13. *Market Forces* dominates until its internal contradictions lead to a global crisis, as *Fortress World* forces surge briefly and ineffectually. *Policy Reform* ascends in the wake of the crisis. Eventually the *Great Transition* era begins as the long-brewing popular desire for fundamental change surges.

The analysis suggests that the momentum toward an unsustainable future can be reversed, but only with great difficulty. The *Great Transition* assumes fundamental shifts in desired lifestyles, values and technology. Yet, even under these assumptions, it takes many decades to realign human activity with a healthy environment,

Figure 13. Overlapping Tendencies in a *Great Transition*

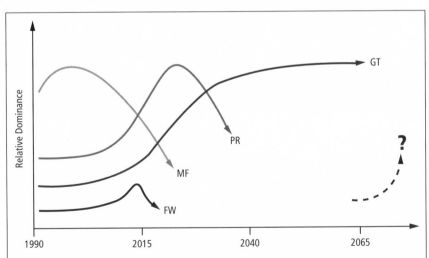

MF = *Market Forces*, PR = *Policy Reform*, GT = *Great Transition*, FW = *Fortress World*

make poverty obsolete, and ameliorate the deep fissures that divide people. Some climate change is irrevocable, water stress will persist in many places, extinct species will not return, and lives will be lost to deprivation.

Nevertheless, a planetary transition toward a humane, just and ecological future is possible. But the curve of development must be bent twice. A radical revision of technological means begins the transition. A reconsideration of human goals completes it. This is the promise and the lure of the global future.

References

Banuri, T., E. Spanger-Siegfried, K. Saeed and S. Waddell. 2001. *Global Public Policy Networks: An emerging innovation in policy development and application*. Boston: SEI-B/Tellus Institute.

Barber, B. 1995. *Jihad vs. McWorld*. New York: Random House.

Bossel, H. 1998. *Earth at a Crossroads. Paths to a Sustainable Future*. Cambridge, UK: Cambridge University Press.

BSD (Board on Sustainable Development of the U.S. National Research Council). 1998. *Our Common Journey: Navigating a Sustainability Transition*. Washington, D.C.: National Academy Press.

CBD (Convention on Biological Diversity). 2001. *See http://www.biodiv.org/*.

CCD (Convention to Combat Desertification). 2001. *See http://www.unccd.int*.

Dominguez, J. and V. Robin. 1992. *Your Money or Your Life*. NY: Viking Penguin.

ECI (Earth Charter Initiative). 2000. *The Earth Charter*. San José, Costa Rica: Earth Charter Commission Secretariat. *See http://www.earthcharter.org /draft/charter.rtf*.

Ehrlich, P. 1968. *The Population Bomb*. NY: Ballantine.

FAO (Food and Agricultural Organization). 1996. *Rome Declaration on World Food Security and World Food Summit Plan of Action. http://www.fao.org*.

Ferguson, N. (ed.) 1999. *Virtual History: Alternatives and Counterfactuals*. NY: Basic Books.

Ferrer, A. 1996. *Historia de la Globalización: orígenes del orden económico mundial*. México: Fondo de Cultura Económica.

Florini, A. 2000. *The Third Force: The Rise of Transnational Civil Society*. NY: Carnegie Endowment.

Gallopín, G. A. Hammond, P. Raskin and R. Swart. 1997. *Branch Points: Global Scenarios and Human Choice*. Stockholm, Sweden: Stockholm Environment Institute. PoleStar Series Report No. 7. *See http://www.gsg.org*.

Gandhi, M. 1993. *The Essential Writings of Mahatma Gandhi*. NY: Oxford University Press.

Harris, P. 1992. *The Third Revolution*. London: Tauris.

Held, H., A. McGrew, D. Goldblatt and J. Perraton. 1999. *Global Transformations: Politics, Economics and Culture*. Stanford, CA: Stanford University Press.

Helleiner, G. 2000. "Markets, Politics and Globalization: Can The Global Economy Be Civilized?" Tenth Raúl Prebisch Lecture, Geneva, 11 December.

Hobbes, T. (1651). 1977. *The Leviathan*. NY: Penguin.

IPCC (International Panel on Climate Change). 2001. *Climate Change 2001: Impacts, Adaptation and Vulnerability* (McCarthy, J., O. Canziani, N. Leary, D. Dokken and K. White, eds.). Cambridge, UK: Cambridge University Press.

Kaplan, R. 2000. *The Coming Anarchy*. NY: Random House.

Kates, R., W. Clark, R. Corell, J. Hall, C. Jaeger, I. Lowe, J. McCarthy, H. Schellnhuber, B. Bolin, N. Dickson, S. Faucheux, G. Gallopín, A. Gruebler, B. Huntley, J. Jäger, N. Jodha, R. Kasperson, A. Mabogunje, P. Matson, H. Mooney, B. Moore, T. O'Riordan and U. Svedin. 2001. "Sustainability science." *Science* 292: 641–642.

Kemp-Benedict, E., C. Heaps and P. Raskin. 2002. *Global Scenario Group Futures: Technical Notes*. Boston: Stockholm Environment Institute-Boston. *See http://www.gsg.org.*

Keynes, J. M. 1936. *The General Theory of Employment, Interest, and Money*. London: MacMillan.

Keynes, J. M. 1972 (first published 1930). "Economic Possibilities for our Grandchildren," in *The Collected Writings of John Maynard Keynes. Vol. IX: Essays and Persuasions*. London: MacMillan.

Lindblom, C. 1959. "The science of 'Muddling Through'" *Public Administration Review* XIX: 79–89.

Maddison, A. 1991. *Dynamic Forces in Capitalist Development. A Long-Run Comparative View*. Oxford: Oxford University Press.

Malthus, T. (1798). 1983. *An Essay on the Principle of Population*. U.S.: Penguin.

Martens, P. and J. Rotmans (eds.). 2001. *Transitions in a Globalising World*. Maastricht: ICIS (manuscript).

Maslow, A. 1954. *Motivation and Personality*. New York: Harper Brothers.

Meadows, D. H., D. L. Meadows, J. Randers and W. W. Behrens. 1972. *Limits to Growth*. New York: Universe Books.

Mill, J. S. (1848). 1998. *Principles of Political Economy*. Oxford, UK: Oxford University Press.

Munasinghe, M. 1999. "Development, Equity and Sustainability in the Context of Climate Change," in: *Climate Change and its Linkages with Development, Equity and Sustainability* (M. Munasinghe and R. Swarts, eds.). Washington, D.C.: LIFE/RIVM/World Bank.

PoleStar (The PoleStar System). 2000. SEI-Boston *See http://www.tellus.org/ seib/publications/ps2000.pdf.*

Raskin, P., G. Gallopín, P. Gutman, A. Hammond and R. Swart 1998. *Bending the Curve: Toward Global Sustainability.* Stockholm, Sweden: Stockholm Environment Institute. PoleStar Series Report No. 8. See *http://www.gsg.org.*

Reinicke, W., F. Deng, T. Benner, J. Gershman and B. Whitaker (eds.). 2000. *Critical Choices: The United Nations, Networks, and the Future of Global Governance.* Ottawa: IDRC.

Renner, M. 1994. *Budgeting for Disarmament: the Costs of War and Peace.* Worldwatch paper 122. Washington D.C.: Worldwatch.

Robinson, J. and J. Tinker. 1996. *Reconciling Ecological, Economic and Social Imperatives: Towards an Analytical Framework.* Vancouver: Sustainable Development Research Institute (UBC).

Sales, K. 2000. *Dwellers in the Land. The Bioregional Vision.* Athens, GA: University of Georgia Press.

Schumacher, E. F. 1972. *Small is Beautiful.* London: Blond and Briggs.

Smith, A. (1776). 1991. *The Wealth of Nations.* Amherst, NY: Prometheus.

Speth, G. 1992. "The transition to a sustainable society," *Proc. Natl. Acad. Sci. USA* 89: 870–872.

Sunkel, O. 2001. "La Sostenibilidad del Desarrollo Vigente en America Latina"; in *Comisión Sudamericana de Paz, Seguridad y Democracia. América Latina en el siglo XXI. De la esperanza a la equidad.* Mexico D.F.: Fondo de Cultura Económica.

Thompson, P. 1993. *The Work of William Morris.* Oxford: Oxford University Press.

UNDP (United Nations Development Program). 2001. *Human Development Report* 2000. Oxford: Oxford University Press.

UNFCCC (United Nations Framework Convention on Climate Change). 1997. *Kyoto Protocol to the United Nations Framework Convention on Climate Change.* See *http://www.unfccc.de.*

UNPD (United Nations Population Division). 2001. *World Population Prospect. The 2000 Revision. Highlights.* NY: United Nations.

Van Parijs, P. 2000. "A Basic Income for All." *Boston Review.* See *http://bostonreview.mit.edu/BR25.5/vanparijs.html.*

Watson, R. T., J. A. Dixon, S. P. Hamburg, A. C. Janetos and R. H. Moss. 1998. *Protecting Our Planet, Securing our Future: Linkages Among Global Environmental Issues and Human Needs.* Washington, D.C.: UNEP/USNASA/World Bank.

WCED (World Commission on Environment and Development). 1987. *Our Common Future.* Oxford: Oxford University Press.